The General Kelvin-Helmholtz Stability Model

Hermann Moshagen

The General Kelvin-Helmholtz Stability Model

With Applications to Sand Wave and Water Wave Generation

Springer

Hermann Moshagen
Norwegian University of Science and Technology
Trondheim, Norway

ISBN 978-3-031-61983-0 ISBN 978-3-031-61984-7 (eBook)
https://doi.org/10.1007/978-3-031-61984-7

This Springer imprint is published by the registered company Springer Nature Switzerland AG
The registered company address is: Gewerbestrasse 11, 6330 Cham, Switzerland

To my wife Eli

Preface

The main lines of thinking in this book go back to my master's degree work in the 1960s. My research work at that time was on stratified flow, supervised by Professor Einar Høiland, hydrodynamics professor at the Institute of Mathematics at the University of Oslo. My work included a study of the book by Professor Chia-Shun Yih: *Dynamics of Nonhomogeneous Fluids*, published in 1965. A second edition came in 1980 with the title *Stratified Flows*. My Master thesis research work was performed at the River and Harbor Laboratory in Trondheim. During this period, I had many stimulating discussions on stratified flows with Dr. Torkild Carstens, Professor of Civil Engineering at NTH (Norwegian Technical Highschool, now NTNU, the Norwegian University of Science and Technology) in Trondheim. I found the problems and methods of stratified flow highly interesting and fascinating.

When I later selected to study the generation of sand waves in alluvial channels as the subject of my research work for the degree of Doctor of Engineering, my approach was again to apply the concepts and methods of stratified flow. The General Kelvin–Helmholtz (GKH) stability model was developed as part of my work. This work took place at the Department of Civil and Environmental Engineering at NTNU, where Professor Geir Moe had the overall responsibility of the work. The results from my research work were not published at the time when the work was completed and has remained unpublished since. However, my recent work has given me good reasons to believe that my results are still relevant for publication. A good part of this book is thus based on my research work in the early 1980s.

During my work on this book, I saw good arguments for including also another limited, introductory, and more recent piece of work on a different subject, namely that of the generation of water waves by wind. The study leads to results that seem to deserve interest, being another application of the GKH stability model.

I have had kind support from several individuals during my work with this book. Engineering manager, Mr. Gudmund Vagnildhaug, of the engineering company Reinertsen AS in Trondheim, provided office space and other practical support during the initial phases of my work. My friend and colleague over many years,

Mr. Knut Reed, has produced most of the diagrams. Another friend, Professor Thomas McClimans, NTNU, reviewed my first manuscript and gave many highly valid comments and suggestions.

My deeply felt gratitude goes to the people now mentioned. Through the engagement and encouragement of the professors I worked with, my knowledge and fascination for stratified flow were developed. Due to the backing on practical matters as well as comments and suggestions on the content, this book was realized. My gratitude is extended to editors Nemul Khan and Tina Silva of Springer for their excellent cooperation in producing the book.

The book is written for and, I hope, will be read by a wide group of readers who do not necessarily have a background in applied mathematics and hydrodynamic stability theory. The reason is that the results are relevant for both engineers, oceanographers, and geologists (sand wave generation), in addition to scientists of fluid mechanics. The mathematics in the book is on a level that should not frighten anybody from reading it. Having this in mind, an introductory section on the character of hydrodynamic stability analysis is included. It is my hope that the results in the book will be considered useful and that the General Kelvin-Helmholtz stability model will be found interesting and valid for use in future scientific and engineering work. The book is organized in three main chapters. These chapters are self standing parts of the book having an introductory abstract and a list of references at the end.

Trondheim, Norway
April 2024

Hermann Moshagen

Contents

Chapter 1
The General Kelvin–Helmholtz Stability Model

1 Hydrodynamic Stability

1.1 Character of Hydrodynamic Stability Work

The hydrodynamic stability problem is about whether a basic steady state flow will remain in its steady state or not. If the flow remains in its steady state, the basic flow is said to be stable. If the basic flow, by the physical conditions governing it, changes to a different flow state, the basic flow is said to be unstable. Thus, stability analysis gives a result characterizing the basic flow. The solution will normally describe the stability mechanism of the flow and explain under what conditions it gets unstable. The main objectives of a hydrodynamic stability study may be stated as follows:

1. To establish a criterion for stability of the basic flow. The criterion shall preferably be formulated in terms of main flow parameters that can readily be established.
2. To model the flow system in such a way that it, together with the stability criterion, provides a good basis for understanding:
   (a) Why the flow will remain stable or not, and
   (b) When unstable, why it develops as it does.

Theoretical stability analysis follows a generally agreed approach where a flow system is defined by a basic flow in terms of a velocity field plus an equation expressing continuity of the flow as well as a relation for the physical characteristics of the fluid. The basic flow is given a small wave-like perturbation by

$$\eta_j = a_j \exp\ i(kx - \sigma t) \tag{1.1}$$

Here k is the wave number ($k = 2\pi/\lambda$, where λ is the wavelength) and σ is the wave frequency ($\sigma = 2\pi/T$, where T is the wave period). Normally, the wave

H. Moshagen, *The General Kelvin-Helmholtz Stability Model*,
https://doi.org/10.1007/978-3-031-61984-7_1

frequency may be replaced by the phase velocity c by the formula $\sigma = ck$. Hereby we may allow the alternative writing of the interface perturbation by:

$$\eta_j = a_j \exp\ ik(x - ct) \tag{1.2}$$

Thus, the stability of a system may be discussed in terms of the phase velocity or in terms of the wave frequency of the perturbation. The stability discussions will mainly be in terms of the phase velocity in the remaining chapter of this book. We should keep in mind, however, that σ gives the real growth rate of the perturbation.

A stability analysis result will describe an initial flow behavior starting from the basic flow. In the case of a stable basic flow, the disturbances will not grow, leaving the basic flow undisturbed. When the basic flow is unstable, the disturbances will grow exponentially and develop the basic flow into a new main flow. Calculating the final resulting flow starting by a hydrodynamic instability will normally be a quite different task, requiring a mathematical approach accounting for all relevant non-linearities in the hydrodynamic equations.

It is seen that, in theoretical approaches, premises like the following are applied:

- Empirical data defining the basic flow
- Empirical data defining the geometric configuration of the system, for Example, a sinusoidal bed, when relevant for the actual problem.

Hydrodynamic stability analysis is traditionally a discipline of applied mathematics. Refined mathematical methods are often used, and strong theoretical efforts are often necessary. However, when the above two objectives are met, and when high-quality physical data from relevant laboratory tests, or from full scale measurements, confirm the derived theoretical results, we then have a good and useful piece of hydrodynamic work at hand. Thus, proper theoretical work and good physical data from measurements, in the laboratory or in the field, are equally valued parts of the same research programs on hydrodynamic stability.

1.2 The KH Instability in the Literature

The Kelvin–Helmholtz instability was first discovered and described by the German scientist Hermann von Helmholtz in 1868 [1]. He described that the interface between two deep inviscid and homogeneous fluid layers of the same density sliding relative to each other, that is, separated by a vortex sheet, will be unstable. This means that a disturbance on the interface will grow in amplitude and eventually generate turbulence at the interface. This instability was later described independently and formulated mathematically by William Thomson (later Lord Kelvin) in 1871 [2].

The KH instability has been further developed—or generalized—to allow the two layers to have finite depth and different densities. The density difference allows the gravity field to stabilize the interface for a range of wavelengths. As will be shown, an interface tension may also be included.

The literature on hydrodynamic stability is extensive. In H. Lamb's book [3], the mathematical formulation as well as the stability result for the conventional Kelvin–Helmholtz model is presented, with a separate section when including the interface tension.

S. Chandrasekhar [4], in his book on hydrodynamic and hydromagnetic stability, discusses the KH instability studying various physical effects, including the effect of surface tension and the effect of rotation. Chandrasekar also considers the development of the KH instability when a transition layer exists between the two main layers.

C. S. Yih, in his book on stratified flow [5], presents, including a clear mathematical formulation, the conventional KH stability model—two layers of different density with one vortex sheet—as an introductory part of his chapter on hydrodynamic instability.

The book by P. G. Drazin and W. H. Reid [6], is an updated and recognized reference on hydrodynamic stability work. The authors present the basic concepts of stability theory, cover the classical stability theory, and present the main results within thermal instability, centrifugal instability, as well as viscous and inviscid theoretical results for parallel shear flows. Drazin and Reid discuss the conventional Kelvin–Helmholtz (KH) instability for homogeneous and stratified flows.

A. D. D. Craik, in his book *Wave Interactions and Fluid Flows* [7], also discusses the KH instability. He points out that fixed boundaries normally are stabilizing the KH instability by decreasing the growth rate. Craik has an interesting discussion of "explosive" behavior of wave modes. This behavior is the result of resonance between the wave modes, as different from a hydrodynamic instability.

Quite some studies have been performed over the recent decades on developing computer models showing the further development of an unstable perturbation, the instability being caused by the KH mechanism. Such models must include the nonlinear terms of the flow of the boundary conditions. The work by Lee and Kim [8], is one example. They show a nice development of the curls following from the instability for flow systems with two and three layers.

An overwhelming number of articles are published during the last decades on KH instability in plasma physics and fluid cases applicable in astrophysics and space research. Although they are of high interest within their fields, they will not be considered further here.

An impression from the literature on the KH instability is that the understanding of it is the conventional stability model as described in the above books. This means that the widespread understanding of the KH mechanism is the instability generated by two layers of flow, possibly having different densities, and separated by a vortex sheet.

Attempts that are made over the years on applying the KH instability to explain and describe important instabilities that are known to exist have given limited success. Typical examples here are the generation of sand waves in alluvial channels and the generation of water waves by wind. This situation appears as a bit of a mismatch. The KH instability is given a closer look in the following.

2 The Kelvin–Helmholtz Instability

2.1 Theoretical Flow Model

Figure 1.1 illustrates a flow system where the KH instability will be in action. An upper fluid layer of large depth, density ρ_1, and velocity U_1 is moving on top of a lower fluid layer also of large depth but with larger density ρ_2 and velocity U_2. The two layers are assumed inviscid and are further assumed to be separated by a vortex sheet of zero thickness. The system is defined in a gravity field and an interface tension T_s is allowed on the interface.

The stability of the interface is analyzed mathematically by introducing a perturbation of the system by allowing a small amplitude wave $\eta = a \exp ik(x - ct)$ on the interface as shown in Fig. 1.1 below.

The mathematical formulation of the conventional KH instability is not included here as it is assumed to be well known. The approach is by flow potentials for the disturbed flow in both layers and by linearized kinematic and dynamic boundary conditions at the interface. The perturbation of the interface is assumed small enough to allow the interface stability to be studied by linearized boundary conditions.

2.2 Stability Analysis Result

The resulting equation gives the phase velocity c of the perturbation on the interface:

$$
\begin{aligned}
c = \sigma/k = &(\rho_1 U_1 + \rho_2 U_2)/(\rho_1 + \rho_2) \\
&+ \left\{\Delta\rho\, g/k/(\rho_1 + \rho_2) + T_s k/(\rho_1 + \rho_2) - \rho_1 \rho_2\, \Delta U^2/(\rho_1 + \rho_2)^2\right\}^{1/2}
\end{aligned}
\tag{1.3}
$$

Referring to Fig. 1.1, U_1, U_2, and ρ_1, ρ_2 are the velocities and densities of the two layers while g is the acceleration of gravity and T_s is the interface tension. The symbols $\Delta\rho$ and ΔU are the differences in density and velocity between the two layers.

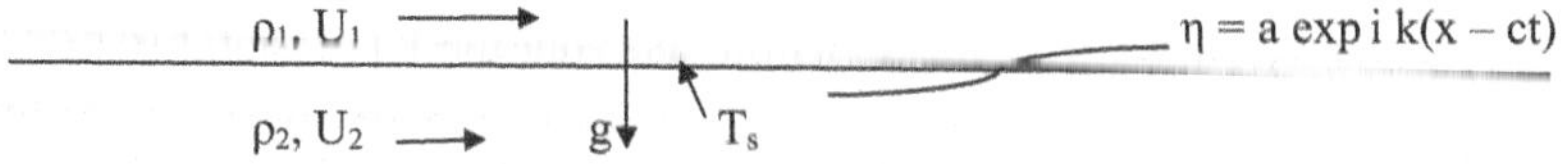

Fig. 1.1 Deep perturbed 2-layer system where the KH instability will develop

According to this equation, the interface is unstable, caused by the velocity difference. We see from the equation that the gravity stabilizes long perturbations while the interface tension T_s stabilizes short perturbations. Thus, the interface is unstable for all wavelengths in the intermediate range between the stable long and short waves. While the interface tension T_s plays an important role for short wave lengths of the disturbance, it has no significant effect for longer waves. To limit discussions, the interface tension will be omitted in the remaining part of the book. When of interest, it can easily be included. Generalizing the above result to a two-layer system with finite depths is straight forward. This is shown by Lamb [3], and is also shown by the example in Sect. 4.3.

The present understanding of the KH mechanism, as stated above, calls for a generalized KH model for flow systems with more flow layers. This generalized KH stability model will be considered in the following.

3 K-H Stability Model for a Multilayer System

3.1 Theoretical Flow Model

A flow system of n fluid layers, having n–1 interfaces, is considered. The layers are superposed on each other in the gravity field and the system is assumed to be inviscid, incompressible, and statically stable. Layer j is given a constant density ρ_j, a thickness h_j, and a constant velocity U_j, as shown in Fig. 1.2. The system is assumed limited by a fixed bed below and fixed surface at the top, without loss of generality.

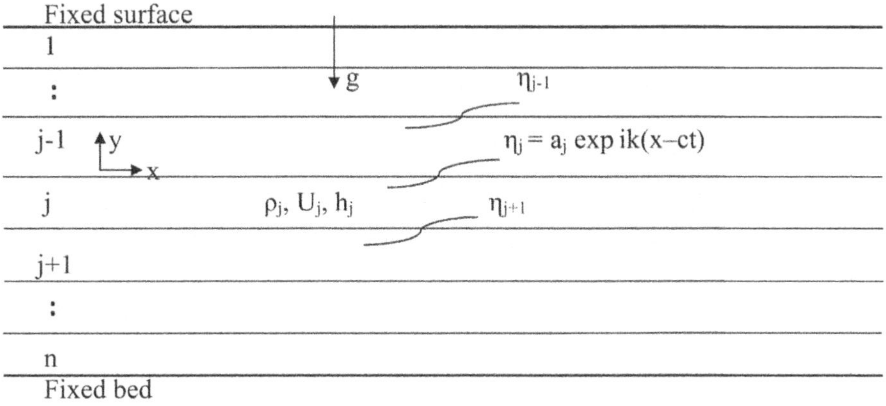

Fig. 1.2 Theoretical flow model of n discrete fluid layers in a gravity field

3.2 Theoretical Solution

A coordinate system is introduced at interface no. j. The flow system is given a perturbation $\eta_j = a_j \exp ik(x - ct)$, $j = 1, 2, \ldots, n - 1$. Different amplitudes on the interfaces, and a phase shift between them, are allowed. Here c is the phase velocity of the perturbation, defined by $c = \sigma/k$. The mathematical approach, for studying the KH instability of this system, is parallel to the one applied for the traditional KH instability, as described in, for example, Yih [5]. We consider interface no. j with its two neighboring layers j below and j–1 above. The perturbed flow is assumed to be irrotational and described by velocity potentials Φ_j. The velocity potentials satisfy Laplace's equation:

$$\nabla^2\Phi_j = 0, \quad j = 1, 2, \ldots, n \tag{1.4}$$

The velocity potential for layer j is formulated as follows:

$$\Phi_j = \left\{C_{j,1} \sinh\, k\left(y - h_j\right) + C_{j,2} \cosh\, k\left(y - h_j\right)\right\} \exp\, ik(x - ct) \tag{1.5}$$

The integration constants, $C_{j,1}$ and $C_{j,2}$, are determined by the linearized kinematic condition at interfaces j and j + 1:

$$\partial\eta_j/\partial t + U_j\partial\eta_j/\partial x = -\partial\Phi_j/\partial y \quad \text{at } y = 0 \tag{1.6}$$

$$\partial\eta_{j+1}/\partial t + U_j\partial\eta_{j+1}/\partial x = -\partial\Phi_j/\partial y \quad \text{at } y = -h_j \tag{1.7}$$

Straight forward development gives the following values of the integration constants:

$$C_{j,1} = i\, a_j\left(c - U_j\right) \tag{1.8}$$

$$C_{j,2} = i\left(a_j/\tanh\, kh_j - a_{j+1}/\sinh\, kh_j\right)\left(c - U_j\right) \tag{1.9}$$

The following is velocity potential for layer j:

$$\begin{aligned}\Phi_j = i\left(c - U_j\right)\left\{a_j \sinh\, k\left(y + h_j\right) + \left(a_j/\sinh\, kh_j - a_{j+1}/\sinh\, kh_j\right)\right. \\ \left.\cosh\, k\left(y + h_{j-1}\right)\right\} \exp\, ik(x - ct)\end{aligned} \tag{1.10}$$

A similar velocity potential applies for layer j − 1:

$$\Phi_{j-1} = i(c - U_{j-1})\{a_{j-1} \sinh k(y + h_{j-1}) + (a_{j-1}/\sinh kh_{j-1} - a_j/\sinh kh_{j-1}) \cosh k(y + h_{j-1})\} \exp ik(x - ct) \tag{1.11}$$

The above potentials for layers j and j − 1 are now applied when entering the dynamic boundary condition of interface j, requiring continuity of pressure across the interface. The linearized dynamic boundary condition reads as follows:

$$\begin{aligned} &\rho_{j-1}\{\partial\Phi_{j-1}/\partial t + U_{j-1}\partial\Phi_{j-1}/\partial x - gy\} \\ &= \rho_j\{\partial\Phi_j/\partial t + U_j\partial\Phi_j/\partial x - gy\} \quad \text{at } y = \eta_j \end{aligned} \tag{1.12}$$

After substitution of potentials and some rearrangement of terms, we arrive at the eigenvalue equation, in terms of the phase velocity c, for interface j:

$$\begin{aligned} &\rho_{j-1}(1/\tanh kh_{j-1} - a_{j-1,j}/\sinh kh_{j-1})(c - U_{j-1})^2 \\ &\quad + \rho_j(1/\tanh kh_j - a_{j+1,j}/\sinh kh_j)(c - U_j)^2 = (\rho_j - \rho_{j-1})g/k, \\ &j = 1, 2, \ldots, n - 1 \end{aligned} \tag{1.13}$$

Here the symbol $a_{j-1,j}$ is a short notation for the amplitude ratio a_{j-1}/a_j and similarly for $a_{j+1,j}$.

3.3 *Comments to the Solution*

The result (1.13) is an eigenvalue equation of Kelvin–Helmholtz type, including now the three amplitudes. This means that the stabilty of interface j is influenced by the neigboring faces j−1 and j+1. It applies to any interface, or a free surface, in a flow system of n discrete layers. A free surface at level j is represented by the requirement $\rho_{j-1} = 0$. The General Kelvin–Helmholtz (GKH) model (1.13) was established by Moshagen [9].

The result (1.13) is clearly not a stability requirement by itself, containing the three amplitudes. However, it may be taken as the starting point for investigation of KH type of instability for any discretely layered flow system once the flow system is specified.

To show that the above GKH resulting equation is consistent with established results, we include some consistency checks on it in the following. Finally, we apply the GKH model (1.13) to a two-layer channel flow having a free surface and one interface, to show how the model applies to this system.

4 Consistency Checks

4.1 Stationary n-Layer System

A first check on the above eigenvalue equation (1.13) is made for a stationary n-layer system, with n − 1 interfaces and with fixed surface and fixed bed. We set the velocity equal to zero in all layers and apply the above equation to all n − 1 interfaces. We rewrite the GKH equation as follows:

$$\rho_{j-1}\left(a_j/\tanh kh_{j-1} - a_{j-1}/\sinh kh_{j-1}\right)c^2 + \rho_j\left(a_j/\tanh kh_j - a_{j+1}/\sinh kh_j\right)c^2 = a_j\left(\rho_j - \rho_{j-1}\right)g/k \tag{1.14}$$

or

$$-\left\{\rho_{j-1}\,kc^2/\sinh kh_{j-1}\right\}a_{j-1} + \left\{\rho_{j-1}\,kc^2/\tanh kh_{j-1} + \rho_j\,kc^2/\tanh kh_j - \left(\rho_j - \rho_{j-1}\right)g\right\}a_j - \left\{\rho_j kc^2/\sinh kh_j)\right\}a_{j+1} = 0, \quad j = 1, 2, \ldots, n-1 \tag{1.15}$$

The eigenvalues of this system are now determined by requiring the determinant of the set of n − 1 linear homogeneous amplitude equations for the total system to be zero. Hereby we get the determinant equation:

$$\begin{vmatrix} A_1 & -B_2 & 0 & & 0 \\ -B_2 & A_2 & -B_3 & & \\ & -B_3 & A_3 & & \\ \cdots & \cdots & \cdots & \cdots & \cdots \\ & & & & -B_{n-1} \\ 0 & & & -B_{n-1} & A_{n-1} \end{vmatrix} = 0 \tag{1.16}$$

The coefficients of the determinant are defined by:

$$A_j = \left(\rho_{j-1}/\tanh kh_{j-1} + \rho_j/\tanh kh_j\right)kc^2 - \left(\rho_j - \rho_{j-1}\right)g \tag{1.17}$$

$$B_j = \rho_{j-1}kc^2/\sinh kh_{j-1}, \quad j = 1, 2, \ldots, n-1, \tag{1.18}$$

keeping in mind that $a_{j-1} = 0$ at the fixed top and $a_{j+1} = 0$ at the fixed bed. This is the same n-layer determinant equation giving the eigenvalues of the n-layer system at rest as established by R. R. Webb, see Lamb [3, p. 372].

4.2 Deep Free Surface Flow

A second consistency check is performed on a well-known result for a deep flow with a free surface. The system is shown in Fig. 1.3 below, where we introduce a perturbation $\eta = a \exp ik(x - ct)$ on the free surface.

Applying Eq. (1.13) to this one-layer system, we set $\rho_0 = 0$ and $\rho_1 = \rho$ and assume kh to be a large number. This implies the approximations $\tanh kh = 1$ and $\sinh kh = \infty$. The GKH equation (1.13) gives the simple eigenvalue result:

$$\rho(c-U)^2 = \rho\, g/k \quad \text{or} \quad c = U \pm \sqrt{g/k} \tag{1.19}$$

which is the equation governing the phase velocity of deep-water free surface gravity waves propagating with (+) and against (−) the flow on the free surface.

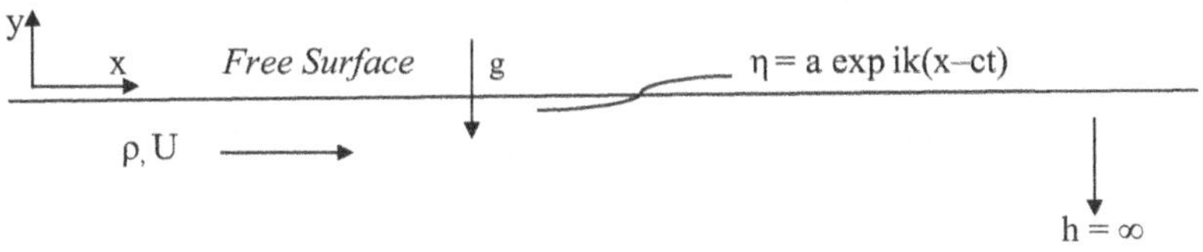

Fig. 1.3 Perturbed free surface deep-water flow

4.3 Two Layers of Finite Depth with Fixed Surface

Another case is considered to check the consistency of the GKH stability model (1.13), focusing on the finite depth two-layer system shown in Fig. 1.4. The system is defined by two inviscid fluid layers of separate densities ρ_1 and ρ_2. The layers are flowing horizontally with constant velocities, U_1 and U_2, respectively. Both layers have finite depths h_1 and h_2. The system is limited by a fixed surface at $y = h_1$ and a fixed bed at $y = -h_2$.

The two layers are separated by an interface at the level $y = 0$. This interface is present as a vortex sheet in the flow system as well as a separating sheet of zero thickness in the density field.

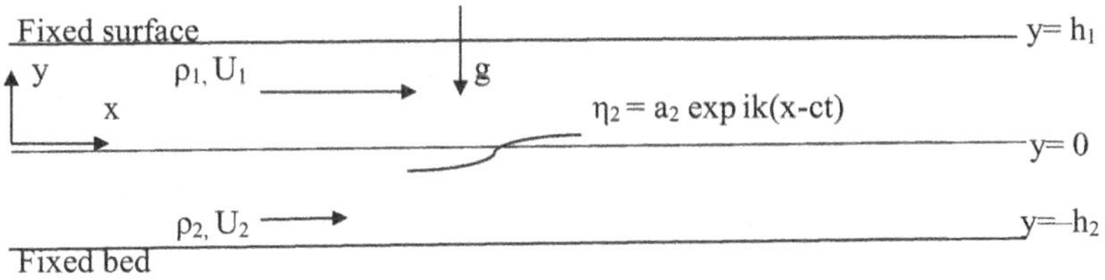

Fig. 1.4 Two-layer system with finite depths. Fixed bed and fixed surface

Here the amplitudes at the surface and at the bed are both zero. Thus, Eq. (1.13) applied to the interface gives us directly the well-known eigenvalue equation:

$$\rho_1{}^*(c - U_1)^2 + \rho_2{}^*(c - U_2)^2 = (\rho_2 - \rho_1)g/k \tag{1.20}$$

The definitions $\rho_1{}^* = \rho_1/\tanh kh_1$ and $\rho_2{}^* = \rho_2/\tanh kh_2$ are applied for shortness in writing. This equation leads to the GKH stability result for this system:

$$\begin{aligned} c = c_r + i\, c_i &= (\rho_1 * U_1 + \rho_2 * U_2)(\rho_1{}^* + \rho_2{}^*)^{-1} \\ &+ i\left\{\rho_1 * \rho_2{}^*(\rho_1{}^* + \rho_2{}^*)^{-2}\Delta U^2 - \Delta\rho(\rho_1{}^* + \rho_2{}^*)^{-1} g/k\right\}^{1/2} \end{aligned} \tag{1.21}$$

This is the well-known KH stability solution for a two-layer system with different densities and with finite depth of the two layers. For a flow system where the depth of the two layers is large compared to the length of all perturbations in question, that is, kh_1 and kh_2, are both large, the hyperbolic tangent functions will be close to 1, and we may set $\rho_1{}^* = \rho_1$ and $\rho_2{}^* = \rho_2$, approximately. We are then back to the conventional KH equation (1.3), disregarding the interface tension T_s.

5 Two Layers of Finite Depth with a Free Surface

We consider, as an application case of the GKH result (1.13), a flow system as illustrated in Fig. 1.5. The system has two inviscid fluid layers of densities ρ_1 and ρ_2, flowing horizontally with constant velocities U_1 and U_2 and with finite depths h_1 and h_2. The system is in the gravity field and is limited by a fixed bed at $y = -h_2$ and a free surface at $y = h_1$.

The GKH equation (1.13) is now applied to the free surface and to the interface, giving the following two equations, respectively:

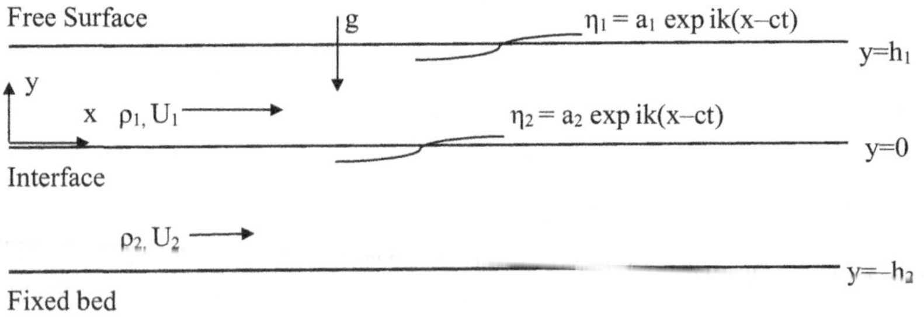

Fig. 1.5 Finite depth two-layer flow system with a free surface

$$\left\{(c-U_1)^2/\tanh kh_1 - g/k\right\}a_1 - \left\{(c-U_1)^2/\sinh kh_1\right\}a_2 = 0 \tag{1.22}$$

$$\left\{-\rho_1(c-U_1)^2/\sinh kh_1\right\}a_1 + \left\{\rho_1{}^*(c-U_1)^2 + \rho_2{}^*(c-U_1)^2 - \Delta\rho g/k\right\}a_2 = 0 \tag{1.23}$$

ρ_1* and ρ_2* are defined as above, see Sect. 4.3. The presence of the amplitudes in the two equations shows that we now have a coupling between the interface and the free surface in our stability solution. The amplitude terms in the two above eigenvalue equations represent an important difference from the conventional KH stability model, Eq. (1.20), for the finite depth system defined in Fig. 1.4, where the surface is fixed. It should be noted from these equations that a coupling between the dynamic faces will take place for perturbations that are long enough while there is no coupling for short perturbations. A stability study, where the two above equations are applied, will be presented in Chap. 2.

6 Concluding Remarks

The General Kelvin–Helmholtz (GKH) stability model is established. It includes the conventional KH stability model as a special case. The GKH stability model applies to all interfaces, and possibly a free surface, in a multilayered flow system. Terms representing a coupling between an interface and the neighboring interfaces, or a free surface, are included in the general model. The GKH model opens for studying the KH stability of flow systems which cannot be studied by the conventional KH stability model.

References

1. von Helmholtz, Hermann. 1868. Über diskontinuierliche Flüssigkeitsbewegungen. *Monatsberichte der Königlich preussischen Akademie der Wissenschaften zu Berlin* 23: 215–228.
2. Thomson, William (Lord Kelvin). 1871. Hydrokinetic solutions and observations. *Philosophical Magazine* 42: 362–377.
3. Lamb, Horace. 1932. *Hydrodynamics*. Cambridge: University Press.
4. Chandrasekar, S. 1961. *Hydrodynamic and Hydromagnetic Stability*. Oxford: Clarendon Press.
5. Yih, Chia-Shun. 1980. *Stratified Flows*. New York: Academic Press.
6. Drazin, P.G., and W.H. Reid. 2004. *Hydrodynamic Stability*. Cambridge: Cambridge University Press.
7. Craik, A.D.D. 1985. *Wave Interactions and Fluid Flows*. Cambridge: Cambridge University Press.

8. Lee, H.G., and J. Kim. 2015. Two-dimensional Kelvin–Helmholtz instabilities of multi-component fluids. *European Journal of Mechanics B/Fluids* 49: 77–88.
9. Moshagen, H. 1984. Sand waves in open channel flows. A study with emphasis on hydrodynamic aspects. Doctoral thesis, NTNU, Trondheim, Norway. https://ntnuopen.ntnu.no/ntnu-xmlui/handle/11250/2622275.

Chapter 2
Stability of a Free Surface Flow with a Bed Layer and Its Relation to Sand Wave Generation: Theoretical Solution and Verification Test

1 Background and Objectives

1.1 Background

Sand waves are met in rivers and in constructed canals, in natural channels and shallow sea areas, and even in deeper offshore areas where the flow and seabed conditions are favorable. Due to their scientific interest and their engineering importance, sand waves have been studied by geologists and by engineers and scientists over several decades. Important studies have been performed and significant progress has been made. The physical mechanism driving their generation from the very beginning on a flat sand bed has been a key question. The present study takes up this question by an approach that is different from earlier work.

1.2 Study Objectives

The following objectives are stated for the present study:

- To develop a hydrodynamic stability theory aiming at giving a contribution to the understanding of the generation and development of the sand waves in alluvial channels.
- To establish a stability criterion defining conditions for generation of bed forms depending on the main characteristics of the flow. The study shall focus on subcritical flow conditions.
- To clarify the influence of the free water surface in the mechanism of generation of the sand waves.
- To clarify the role of the bed layer in the mechanism of generation of the sand waves.

H. Moshagen, *The General Kelvin-Helmholtz Stability Model*,
https://doi.org/10.1007/978-3-031-61984-7_2

The theoretical approach will be by applying the GKH stability result for the multilayer model as established in Chap. 1. The study will be aiming at the generation aspects of the sand waves with emphasis on the waves generated under subcritical flow conditions. Some light will, hopefully, also be shed on the generation and development of bed waves for other flow conditions. The study is limited to two-dimensional flow. This is of course an approximation to the real 3D sand wave generation process in alluvial channels subject to a 3D flow. However, on the positive side, the actual bed features often show a remarkable two-dimensional character. Thus, we may hope that their basic physical character will be represented in a two-dimensional model. The theoretical results shall be compared with observations of sand waves generated in a laboratory test. The main results of the study shall be based on this comparison.

As an introduction to the study, we will first present the classification system for sand waves in alluvial channels as established by ASCE. The ASCE Committee on Sedimentation set up a Task Force on Bed Forms in Alluvial Channels that reported their work on classification of bed forms in 1966 [2]. Their classification is given in the following.

2 Classification of Bed Forms

2.1 ASCE Classification

The Committee on Sedimentation of the American Society of Civil Engineers (ASCE) set up a Task Force on Bed Forms in Alluvial Channels that reported their work on classification of bed forms in 1966 [1]. The various bed forms were characterized as follows:

Bed Form
Any deviation, from a flat bed, that is readily detectable by eye or higher than the largest sediment size present in the parent bed material, generated on the bed of an alluvial channel by the flow.

Ripples
Small bed forms with wave lengths less than approximately 1 ft and heights less than approximately 0.1 ft. Ripples occur at flow velocities slightly higher than those required for incipient sediment motion, but at lower velocities than flat bed or antidunes. Parts of the upstream slopes of dunes are occupied by ripples under some flow conditions. In a plan view, a ripple configuration can vary from an irregular array of three-dimensional peaks and pockets to a regular array of continuous parallel crest and troughs transverse to the direction of flow. In longitudinal section, ripple profiles vary from approximately triangular, with long gentle upstream slopes and downstream slopes approximately equal to the angle of repose of the bed material, to symmetrical nearly sinusoidal shapes. The triangular ripples commonly observed in alluvial channels move downstream with velocities that are small compared to the

mean velocity of the generating flow. They are observed to occur only rarely in sediments coarser than approximately 0.6 mm.

Dunes

Bed forms larger than ripples that are out of phase with any water surface gravity waves that accompany them. Dunes generally occur at larger velocities and sediment transport rates than ripples, but smaller velocities and transport rates than anti-dunes; however, ripples do occur on the upstream slopes of dunes at the lower velocities in the dune regime. The lengths of dune crests are usually of the same order as the wavelength. The longitudinal profiles of dunes are approximately equal to the angle of repose of the bed material. Dunes move slowly downstream with velocities that are small compared to the mean velocity of the generating flow. The large lee eddies that occur in dune troughs often cause surface boils of intense turbulence and high sediment concentration to occur above and slightly downstream from dune crests.

Transition

A bed configuration consisting of a heterogeneous array of bed forms, primarily low-amplitude ripples or dunes and flat areas. The transition bed is a configuration generated by flow conditions intermediate to those producing dunes and flat bed. In laboratory flumes, the transition configuration has been observed in some cases to consist of dunes or ripples over a reach covering part of the channel length and with a sand bed being flat along the remainder. The flow in the flat-bed reach is shallower than in the reach with ripples or dunes, and the discontinuity in the bed between the two reaches moves slowly downstream. In other instances, the bed configuration has been observed to consist of small, widely separated ripples or dunes over the entire bed.

Anti-dunes

Bed forms that occur in trains and that are in phase with and strongly interact with gravity water surface waves. Anti-dunes can move upstream or downstream or can remain stationary, depending on the properties of the flow, fluid, and sediment. The free-surface waves have larger amplitudes than the anti-dunes. At higher velocities and Froude numbers, the surface waves usually grow until they become unstable and break in the upstream direction. The agitation accompanying breaking obliterates the anti-dunes, and the process of anti-dune initiation and growth is then repeated. At lower velocities, the anti-dunes will grow and then diminish in amplitude without the surface waves ever breaking. In longitudinal section, anti-dune profiles vary with flow and sediment properties from approximately triangular to sinusoidal, the latter occurring at larger Froude numbers than the former. However, the sharp-crested, triangular-shaped anti-dunes have been observed only in laboratory flumes and always move downstream. The crest lengths of anti-dunes are usually of the same order as the wavelength.

2.2 Illustration of Bed Forms

The ASCE task force illustrated the different flow states and associated bed forms as shown in the simplified two-dimensional sketches in Fig. 2.1. The illustrations cover the whole range of hydraulic Froude numbers from quite small to well above 1.

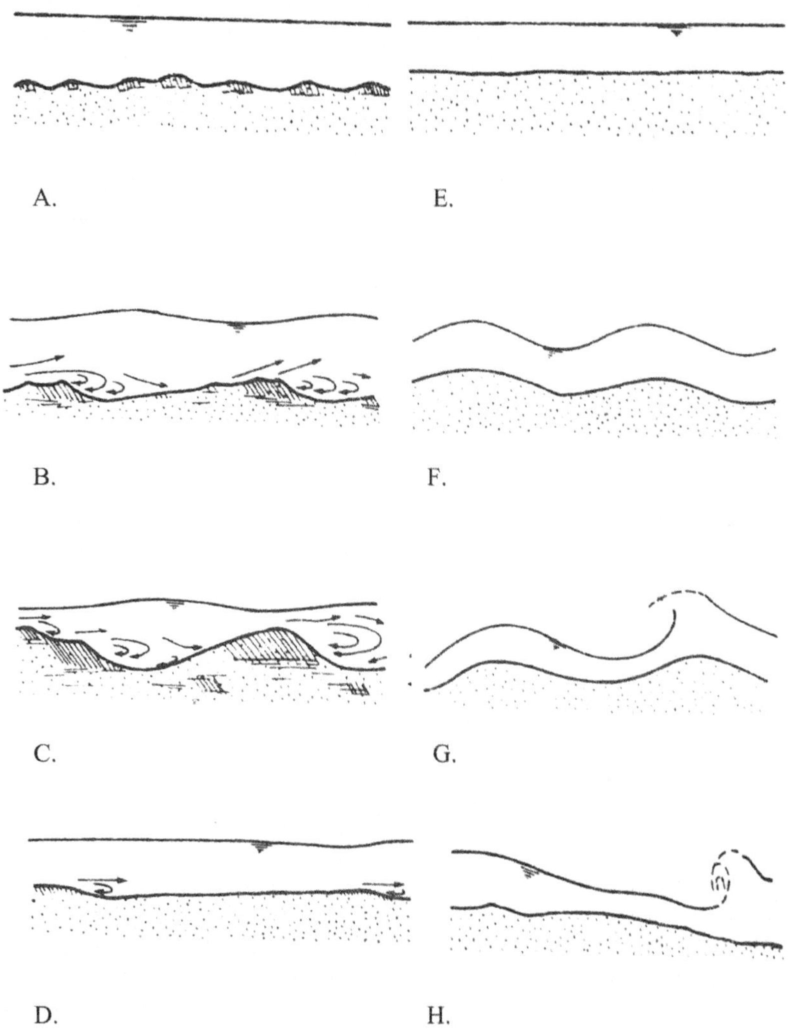

Fig. 2.1 Alluvial channel bed forms and surface wave shape for increasing hydraulic Froude number according to the ASCE classification. (**A**) $F \ll 1$; Ripples, weak flow, surface unaffected, (**B**) $F < 1$; Dunes, ripples on upstream slope, surface out of face, (**C**) $F < 1$; Dunes of different size, surface out of phase, (**D**) $F < 1$; Flow slightly subcritical, dunes washed out, (**E**) $F \approx 1$; Flow slightly supercritical, plane bed or standing waves, surface in phase, (**F**) $F > 1$; Supercritical flow, anti-dunes, surface in phase, (**G**) $F > 1$; Supercritical flow, surface in phase and breaking, (**H**) $F \gg 1$; Flow strongly supercritical, extreme anti-dune forms, chutes, and pools

2.3 Some Comments on the ASCE Classification

The ASCE classification of bed forms is based on agreed visual observations of their appearance and behavior. Ripples and dunes are understood as distinct categories. A consensus exists that ripples are smaller than dunes by their length and height. The maximum length of ripples is proposed to be 1 foot by ASCE, while others propose 60 cm or 1 m. Some authors indicate a transition between ripples and dunes, while others point to the fact that ripples and dunes are seen coexisting for some flow conditions. On the other hand, the development of ripples or dunes over time should fit nicely in here.

We observe the governing role of the Froude number. This indicates that a theoretical model of the generation mechanism in a flow system should be sought by a proper modeling of inertia and gravity. The ASCE classification indicates that a one-to-one correspondence exists between the Froude number and the types of sand waves occurring.

We may note that dunes and anti-dunes both interact with the free surface. The role of the free surface in the generation of the bed waves is, however, not described. The shorter ripples seem to be generated and occur without any interaction with the free surface. It may be added that as dunes and anti-dunes, due to their size, interact strongly with the free surface, they thereby also cause a significant influence on the main flow.

The above generally agreed ASCE classification is established without having a fully valid scientific basis for interpretation of the observations. A good understanding of the dynamics governing the generation of the bed waves would provide such a basis. On this background, the need for an improved scientific basis for generation of the bed wave forms is seen. This is also the background for the many studies that have been performed. In the following, we take a broad review on the analytical work performed on the sand wave generation problem.

3 Existing Theoretical Work

3.1 Review Approach

The bed forms in alluvial channels have been studied over many decades. In the following, we make a brief review on earlier studies on the sand wave problem by addressing the work performed in different periods, thereby illustrating the progress made over the years. The focus will be on the study approaches and the conclusions from the work, while pointing out results of special interest.

3.2 Early Work

Raudkivi [18] refers in his book to publications on the sand wave problem as early as in the eighteenth and the nineteenth centuries as well as in the first part of the twentieth century. Strong theoretical efforts were made from the 1950s. Liu [14] studied the formation of ripples by applying the Kelvin-Helmholtz instability to the bed surface, assuming the bed to behave as a heavy fluid. The approach by considering the sand bed as a fluid was opposed by discussers of the paper.

3.3 A Period of High Activity

Kennedy [13] applied potential theory in describing the basic flow. A sinusoidal configuration was assumed on the sediment bed. Kennedy performed a stability analysis, applying Brooke Benjamins maximum growth rate criterion, to describe the generation and development of dunes and anti-dunes, concluding that his results were in good agreement with experimental data. Reynolds [19] applied channel hydraulics equations and developed further some of Kennedy's results. Raudkivi, in his 1966 paper [16], studies the mechanics of bed wave formation and offers a solution for the relative phase relationship of between the bed wave and the surface wave in alluvial channels. Raudkivi concludes his paper by stating that the development of bed forms from a flat sandy bed is a progressive process and not a spontaneous growth of a bed disturbance.

Jain and Kennedy [12] studied statistically the growth of sand waves observed developing from an initially horizontal bed until their equilibrium configuration. They analyze the sand waves during a progressive development. Jain and Kennedy found that their observations exhibit two distinct peaks. One peak arises from the interaction between the erodible bed and stationery free-surface waves, in agreement with their prediction of accelerated growth near the condition $F^2 = (1/kh) \tanh kh$. The other peak was concluded to be present without a free surface and associated with an instability mechanism generating ripples.

Engelund [9] and Fredsøe [11] applied the vorticity transport equation, leading to an Orr-Sommerfeld type of stability equation. The seabed was given a small amplitude sinusoidal perturbation, subject to a developed flow velocity profile. The eddy viscosity concept as well as an equation expressing sediment conservation was applied. Engelund included internal friction and a non-uniform sediment distribution in his model and concluded that his approach led to results that were rather different from the earlier results based on potential flow analysis. Fredsøe studied the influence of gravity on the bed load transport, finding that this turned out to change the stability results significantly.

Richards [20] studied the influence of the turbulent flow close to the bed. His shear flow model predicted the occurrence of two separate modes of instability. He concluded that these two modes correspond to the formation of ripples and dunes, respectively.

Despite these strong efforts, over several decades, a satisfactory solution to the problem, of how sand waves are generated on a flat bed, was not established by the early 1980s. This is clearly expressed in Engelund's and Fredsøe's [10] status report and in Raudkivi's [17] paper. Professor Raudkivi, in his paper, summarizes the status stating that a satisfactory understanding of the generation and development of sand waves in alluvial channels still has only partial answers.

Thus, the status in the study of the seabed features by the early 1980s was much as follows:

- A high level of detail in water flow and sediment transport are applied premises.
- Various mathematical models as well as a high degree of mathematical refinement have been applied in the theoretical studies.
- All relevant physical parameters seem to be included.

Still, the understanding of the generation bed features, at this stage, was not satisfactory. The situation may indicate that the applied approaches so far had some inherent limitation. Thus, the problem of generation of soft bed features in alluvial channels seemed to call for a different approach. This was the stage of development when my doctoral thesis work on this subject was initiated.

3.4 Later Work

Sumer and Bakioglu [21] extended the theory by Richards in studying the formation of ripples. They concluded that ripples exist when the grain Reynolds number is less than a certain value.

De Jong [8] takes a stratified flow approach to the sand wave problem, applying numerically generated velocity and density profiles. His basic hypothesis is that dunes and ripples are generated by free internal waves in the stratified flow system. One of his main findings is that the presence of bed-load transport has a drastic influence on the behavior of the free internal waves. Comparing his results with experimental data, de Jong concludes that the internal waves may be responsible for the initial generation of the bed waves.

Coleman and Melville [5] focused on the mechanics of bed-form development. They analyzed the results of a series of 47 experiments, covering two types of sand, fine and coarse, and a range of subcritical flow conditions. They worked on data from a progressive development of sand waves and report two important observations:

- The rate of bed-form growth decreases with time as bed forms develop from flatbed conditions.
- An increase in the rate of bed development occurred in the vicinity of the condition $F^2 = (1/kh) \tanh kh$, where k = angular wave number, in line with

the finding by Jain and Kennedy. The period of accelerated growth was flanked by periods of slower development. This development was seen for a series of runs.

Coleman and Melville in their 1996 article [6], based on the same 47 experiments, focused on the generation of initial bed waves, termed wavelets, on a plane bed of fine sand. They determined the wavelength of the initial sand waves and concluded that both ripples and dunes develop from the initial wavelets.

The ASCE Task Committee on Flow and Transport over Dunes summed up the status in 2002 [3]. They stated that in spite of strenuous efforts over the last half century, the engineering prediction of flow and sediment transport over bedforms, in general, and dunes in particular, still presents a major obstacle in the solution of sedimentation problems in alluvial channels.

Colombini [7] makes a revisit to the linear theory of sand dune and anti-dune formation, making an interesting assumption in his theoretical model by evaluating the flow shear at the interface between the main flow and the bed layer and introducing the thickness of the bed layer in his analysis. The "law of the wall" velocity profile, an eddy viscosity, an equation for sediment conservation, and a perturbed seabed are theoretical premises applied in his stability model. Colombini concludes that his model gives a unified description of the generation of dunes and anti-dunes with stability results that seem largely to be in line with those of Fredsøe.

Venditti et al. [22] performed a series of experiments at the National Sedimentation Laboratory of the U. S. Department of Agriculture, using a sand with $D_{50} = 0.5$ mm. They concluded that transverse wave forms in subcritical flow can be initiated as an interfacial Kelvin-Helmholtz instability.

Abedi et al. (2014) reported the existence of interfacial instability in the two-dimensional channel flow of sediment suspension. They showed that a Kelvin-Helmholtz type of unstable wave develops at the interface for wavelengths longer than a critical value.

Ali and Dey [1] apply a one-equation rotational flow model in studying the stability of dunes and anti-dunes. They conclude that their model offers a unique reconstruction of the linear stability map, capturing satisfactorily the experimental data of dunes, anti-dunes, as well as the transition from dunes to anti-dunes.

A significant amount of work has been performed over the later years considering the generation and formation of sand waves and sand banks in shallow sea areas dominated by tidal flows. The work has included development of computer models to simulate the development of bed forms. As the tidal flow is a basic premise in this work, the published papers on these issues are not considered further here.

3.5 *Overall Summary and Notes*

Thus, up to present we have seen a long series of analytical studies attempting to build up an instability solution describing the generation of wave forms on the bed of alluvial channels. An overall agreement exists that the Froude number is the governing parameter in the sand wave generation process. The studies apply different concepts in describing the basic flow, for example, potential flow, equations of channel hydraulics, or vorticity transport equations. Derived flow concepts like eddy viscosity, slip velocity, and others are also applied. Various sediment transport assumptions are included and discussed. The sand bed is normally given an initial small amplitude sinusoidal perturbation, with a migration velocity being the key parameter in the stability analysis, the imaginary part of it expressing the stability character of the bed. Brooke Benjamin's [4] maximum growth rate criterion is applied in several studies in establishing the stability solution.

We note the following points of interest based on the above review of earlier work:

1. The Froude number is agreed to be the governing dynamic parameter for the generation of sand waves in alluvial channels.
2. Venditti et al. as well as Abedi et al.'s finding of a Kelvin-Helmholtz type of unstable wave developing at the interface.
3. Jain and Kennedy's finding, and later Coleman and Melville's confirmation, of accelerated growth near the condition $F^2 = (1/kh) \tanh kh$.
4. Coleman and Melville's observation that the rate of bed-form growth decreases gradually with time as the bed develops sand waves from flatbed conditions.
5. Coleman and Melville observed that the period of accelerated growth was flanked by periods of slower development.
6. Coleman and Melville's observation that both ripples and dunes develop from the initial wavelets.
7. Sand waves develop progressively, as observed by Raudkivi and later studied by Jain and Kennedy as well as by Coleman and Melville.

The overall agreement that the Froude number is the governing dynamic parameter for sand wave generation is seen as indicating that the Kelvin-Helmholtz instability should be checked out more carefully. The two studies under Point 2 above may be seen as strengthening this indication.

4 Study Basis

4.1 Scope of Section

The present section states the following parts of the study basis:

- The most important characteristics of the sand waves
- The main study hypothesis
- The time domain hypothesis
- The premises to be applied in the stability model

4.2 Basic Sand Wave Characteristics

The characteristics of the sand waves, as generally agreed, are stated in the following. They will serve as a background for setting up the theoretical model as well as for the interpretation of the results from the work. With reference to Sect. 2, the following sand wave characteristics are noted:

- The sediment transport for weak flows is, according to generally agreed observations, as bed load only. The sediment grains are transported in a layer along the bed.
- Ripples are observed occurring at flow velocity about or slightly higher than the one required for incipient sediment motion.
- Ripples may be the only bed features, or they may be observed coexisting with dunes.
- Fully developed dunes are longer than the water depth.
- Fully developed dunes have a significant height in comparison to the water depth.
- Dunes are approximately in antiphase with the free surface in two-dimensional flow.
- Ripples are significantly smaller, in both length and height, than the fully developed dunes.
- The growth rate of the bed forms is low, compared to the flow velocity.
- The propagation rate of the bed forms is low, compared to the flow velocity.
- The Froude number is the governing parameter in the generation of bed forms.

These notes, as well as the above notes from earlier work, will be applied in discussing the results from the study.

4.3 Main Study Hypothesis

The Main Study Hypothesis (MSH) is now stated as follows:

Sand waves in alluvial channels are generated by a hydrodynamic instability developed at the shear layer between the main flow and the sediment carrying bed layer. Release and settlement of sediment at the bed is a secondary process, resulting from this hydrodynamic instability.

4.4 Time Domain Hypothesis

We take Raudkivi's observation on the progressive development of the bed waves as our time domain hypothesis (TDH):

The hydrodynamic instability generating sand waves develops progressively, i.e. *short perturbations are generated initially, followed by progressively generated longer perturbations.*

The stability results will primarily be established in the phase velocity domain with supplementary frequency domain considerations, as needed. The TDH will be applied for interpretation of the stability results in the time domain.

4.5 Study Premises

The following premises shall be valid for the theoretical study:

- Inertia and gravity shall be the main physical forces generating bed waves in a free surface flow, i.e., the Froude number shall be the governing hydrodynamic parameter.
- The study shall focus on the subcritical flow regime, i.e., the Froude number shall be well below 1. However, the flow velocity shall be high enough to establish sediment transport in a bed layer.
- The General Kelvin-Helmholtz stability model, as developed in Chap. 1, shall be applied in studying the generation of bed forms in alluvial channels.
- Long waves, in general giving small eigenvalues for the perturbations on the interface between the main flow and the bed layer, shall be focused in the stability analysis.
- The sediment bed shall be modelled as a fixed horizontal bed in the hydrodynamic model, while a release and settlement mechanism for the bed sand shall be considered for the interpretation of the stability results.
- The sediment flow shall be modelled by a sediment carrying bed layer flow and by the sediments' influence on the integrated bed layer density.

The above premises as well as the MSH and the TDH will be considered when discussing the results from the study. Their validity will be considered based on the actual generation process in a physical system as observed in a laboratory test.

The study premises clearly lay the emphasis on the stratified character of the flow while the sediment bed does not to take part in the driving dynamics of the system. The sand transport is included a priori by the presence of the bed layer. The underlying thinking here is, as expressed already:

- The water flow is driving the development of the system.
- The sediment release from the bed and its settlement on the bed are secondary processes resulting from the dynamics of the flow.

4.6 A Final Remark

The present study thus takes a step back from earlier studies by applying a simple two-layer discretized basic flow system for the stability analysis. The sand bed is considered non-dynamic in a hydrodynamic sense. An initial perturbation of the bed surface is therefore not included. By the above study basis, we are now ready to establish the theoretical model of the flow system and to formulate the hydrodynamic equations for the study.

5 Stability Analysis

5.1 Theoretical Flow Model

Based on the above hypotheses and premises, an idealized 2-D flow system will be applied for the study, as shown in Fig. 2.2 below. The main flow is assumed horizontal and of constant velocity, as is the flow in the bed layer. The flow shear is concentrated in vortex sheets at the interface and at the bed surface. The flow system is given a perturbation, at the free surface and at the interface, with a wavelength λ, given by the wave number $k = 2\pi/\lambda$, and with a phase velocity $c = \sigma/k = \lambda/T$. Here $\sigma = 2\pi/T$ is the wave frequency where T is the wave period.

A couple of notes on the applied flow field are made:

- The shear zone between the main flow and the bed layer is modeled as an interface, or mathematically as a vortex sheet, concentrating the effect of the flow shear.
- It should be noted that the interface is not only a vortex sheet in the flow field but is also expressing a discontinuity in the density field, assuming the density to be constant within each layer.

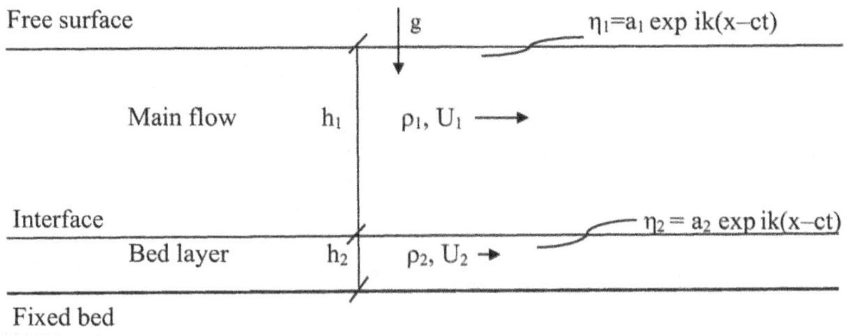

Fig. 2.2 Theoretical flow model of alluvial channel flow with a bed layer

- In an alluvial channel flow, where sediments are part of the flow, a conventional viscous boundary layer concept will not apply. The term *bed layer* will be used throughout the remaining text of this chapter.

It may also be noted that the horizontal uniform and constant velocities U_1 and U_2 are both solutions of Euler's equation of motion. Thus, the theoretical model meets this basic requirement.

5.2 Eigenvalue Equation

In studying the stability of the flow system defined in Fig. 2.2 above, we apply the General Kelvin-Helmholtz (GKH) stability model, Eq. (1.13) of Chap. 1:

$$\begin{aligned}&\rho_{j-1}\left(a_j/\tanh\ k_{j-1}-a_{j-1}/\sinh\ kh_{j-1}\right)\left(c-U_{j-1}\right)^2\\&\quad+\rho_j\left(a_j/\tanh\ kh_j-a_{j+1}/\sinh\ kh_j\right)\left(c-U_j\right)^2=a_j\left(\rho_j-\rho_{j-1}\right)g/k,\\&j=1,2,\ldots,n-1\end{aligned}$$

Applying the GKH equation to the free surface and to the interface, the following two amplitude equations are established, respectively:

$$\left\{(c-U_1)^2/\tanh\ kh_1-g/k\right\}a_1-\left\{(c-U_1)^2/\sinh\ kh_1\right\}a_2=0 \tag{2.1}$$

$$\begin{aligned}&\left\{-\rho_1(c-U_1)^2/\sinh\ kh_1\right\}a_1\\&\quad+\left\{\rho_1{}^*(c-U_1)^2+\rho_2{}^*(c-U_2)^2-\Delta\rho\ g/k\right\}a_2=0\end{aligned} \tag{2.2}$$

The determinant of these two equations is required to be zero, to have a non-trivial solution for the amplitudes. The above two amplitude equations then give the following determinant equation determining the eigenvalues of the system:

$$\begin{aligned}&\left\{\rho_1{}^*(c-U_1)^2-\rho_1\ g/k\right\}\left\{\rho_1{}^*(c-U_1)^2+\rho_2{}^*(c-U_2)^2-\Delta\rho\ g/k\right\}\\&=\rho_1{}^*2(c-U_1)^4/\cosh^2 kh_1\end{aligned} \tag{2.3}$$

The definitions are restated as follows:

- c is the phase velocity while σ and k are the perturbation's frequency and wave number, respectively.
- U_1, U_2 and ρ_1, ρ_2 are the velocities and densities of the two flow layers while g is the acceleration of gravity.

- The symbols $\Delta\rho$ and ΔU are the differences between the densities and velocities of the two layers.
- For shortness in writing, we have defined $\rho_1{}^* = \rho_1/\tanh kh_1$ and similarly for $\rho_2{}^*$.

This equation is now transformed to a new coordinate system, following the flow of the bed layer, by

$$\zeta = c - U_2 \tag{2.4}$$

This implies, splitting in real and imaginary parts, the following understanding of the phase velocity:

$$\zeta_r = c_r - U_2 \quad \text{and} \quad \zeta_i = c_i \tag{2.5}$$

The propagation rate ζ_r is now given in a coordinate system following the flow of the bed layer while the growth rate ζ_i, i.e., the stability solution, is conserved by the transformation of the coordinate system. The transformed equation reads:

$$\begin{aligned} &\left\{\rho_1{}^*(\zeta-\Delta U)^2 - \rho_1\, g/k\right\}\left\{\rho_1{}^*(\zeta-\Delta U)^2 + \rho_2 * \zeta^2 - \Delta\rho g/k\right\} \\ &= \rho_1 *^2(\zeta-\Delta U)^4/\cosh^2 kh_1 \end{aligned} \tag{2.6}$$

Equation (2.6) shows that the velocity difference ΔU is determining for the eigenvalues of the system and not the velocities of the separate layers.

The term on the right represents the coupling between the free surface and the interface at the bed layer. This term is seen to diminish rapidly for large values of the argument kh_1. We will be back to this coupling later in the study.

5.3 Dimensionless Eigenvalue Equation

We give Eq. (2.6) a dimensionless form for eigenvalue analysis by introducing the following definitions:

$$r = \Delta\rho/\rho_1 = (\rho_2 - \rho_1)/\rho_1 \tag{2.7}$$

$$\xi = \zeta/\Delta U = \zeta/(U_1 - U_2) \tag{2.8}$$

$$F^2 = \Delta U^2/gh_1 \tag{2.9}$$

The parameter r is the dimensionless density difference at the bed layer interface while ξ is now the dimensionless phase velocity relative to the bed layer. We note that the Froude number now is defined by the velocity difference and the water depth of the main layer. As the flow velocity of the bed layer is low compared to the velocity of the main flow and the depth of the bed layer is small compared to the depth of the main flow, our hydrodynamic Froude number will be close to, but slightly different from, the conventional Froude number applied in channel hydraulics. In the following, we will discuss the stability of the system in terms of these main flow parameters.

We now re-introduce the definitions of $\rho_1{}^* = \rho_1/\tanh kh_1$ and similarly for ρ_2. The eigenvalue equation (2.6) can now be written as follows:

$$\begin{aligned}&\left\{\tanh kh_1/kh_1 - F^2(\xi-1)^2\right\}(r+1)F^2 \tanh kh_1\, \xi^2\\ &= \left\{F^4 \tanh^2 kh_1 (\xi-1)^4 - (r+1)F^2(\tanh kh_1/kh_1)(\xi-1)^2\right.\\ &\quad \left. + r(\tanh kh_1/kh_1)^2\right\} \tanh kh_2\end{aligned} \tag{2.10}$$

5.4 *Thin Bed Layer Equation*

We develop Eq. (2.10) for application to a system where we have a thin sublayer. We will now search small eigenvalues corresponding to long waves, i.e., the wavelength λ is assumed to be long compared to the depth of the sublayer, as expressed by the requirement $kh_2 \ll 1$. Based on this assumption, we may introduce the approximation

$$\tanh kh_2 = kh_2 + O\left\{(kh_2)^2\right\} \tag{2.11}$$

in Eq. (2.10). Thereby we get the long wave eigenvalue equation for the open channel flow system with a free surface and a thin heavy sublayer at the bed. The following equation now determines the eigenvalues for this system:

$$\begin{aligned}&\left\{\tanh kh_1/kh_1 - F^2(\xi-1)^2\right\}(r+1)F^2 \tanh kh_1 \xi^2\\ &= \left\{F^4 \tanh^2 kh_1 (\xi-1)^4 - (r+1)F^2(\tanh kh_1/kh_1)(\xi-1)^2\right.\\ &\quad \left. + r(\tanh kh_1/kh_1)^2\right\} kh_2 + O\left\{(kh_2)^2\right\}\end{aligned} \tag{2.12}$$

Equation (2.12) determines ξ in terms of the main flow variables F, r, and kh_1, while we will develop the solution in terms of kh_2, assumed small.

6 Stability Solution

6.1 Solution Procedure

The basic assumption in the development of long wave, and normally small, eigenvalues is that $kh_2 \ll 1$. Inspection of Eq. (2.12) reveals that small eigenvalues of ξ will be of order $(kh_2)^{1/2}$. We therefore apply:

$$\varepsilon = (kh_2)^{1/2} \tag{2.13}$$

as the perturbation parameter to be applied in the asymptotic development of the small eigenvalues. Reference is made to the asymptotic development of eigenvalues in Appendix A. Both finite and small eigenvalues are developed for completeness in the Appendix, while the stability solution, discussed in the following, considers the small eigenvalues only.

6.2 Small Eigenvalue Solution

The small eigenvalue solution is established by the asymptotic development in the Appendix and is given by Eq. (A.25) as follows:

$$\begin{aligned} \xi = & i\{F^2 - f_1(kh_1, r)\}^{1/2}\{F^2 - f_2(kh_1, r)\}^{1/2} \\ & \cdot\{(r+1)(F^2 - f_3(kh_1))F^2 \coth kh_1\}^{-1/2}(kh_2)^{1/2} \\ & + \left\{(F^2 - f_3(kh_1))^2 + (1 - \tanh^2 kh_1)(kh_1)^{-2}\right\} \\ & \cdot\left\{(F^2 - f_3(kh_1))^2(r+1)\right\}^{-1} \tanh kh_1\, kh_2 + O\left\{(kh_2)^{3/2}\right\} \end{aligned} \tag{2.14}$$

We include the two lowest order terms for reasons that will appear below. The functions f_1, f_2, and f_3 are defined by the following expressions:

$$\begin{aligned} f_1(kh_1, r) = & \tfrac{1}{2}(r+1)\coth kh_1/kh_1 \\ & + \left\{(r+1)^2(\coth kh_1/kh_1)^2/4 - r/(kh_1)^2\right\}^{1/2} \end{aligned} \tag{2.15}$$

$$\begin{aligned} f_2(kh_1, r) = & \tfrac{1}{2}(r+1)\coth kh_1/kh_1 \\ & - \left\{(r+1)^2(\coth kh_1/kh_1)^2/4 - r/(kh_1)^2\right\}^{1/2} \end{aligned} \tag{2.16}$$

$$f_3(kh_1) = \tanh kh1/kh1 \tag{2.17}$$

By inspection of the above eigenvalue solution (2.14), we see that the second term, $O\{kh_2\}$, is always real and positive, i.e., a propagation rate term. The stability result is given by the first term, $O\{(kh_2)^{1/2}\}$, disregarding higher order terms. The stability limits, at $\xi_i = 0$, are given by the requirements $F^2 = f_1(kh_1, r)$ and $F^2 = f_2(kh_1, r)$.

The third requirement $F^2 = f_3(kh_1)$ is a stability limit in the sense that the growth rate changes sign at it, i.e., it represents a transition from a stable zone on the one side to an unstable zone on the other. However, as this requirement is in the denominator of the growth rate function, it is also a singularity condition for the growth rate function.

The stability zones, as defined by the above stability limits, are illustrated in Figs. 2.3 in Sect. 7 below.

7 Illustration of Stability Results

7.1 *Stability Zones*

Figures 2.3 and 2.4 illustrates the stability zones by their limits. The curves defining the lower limits of the stability Zones II and V are dependent on the value of r. The stability limits are drawn with r as a parameter ranging from 0.25 to 1.00 to illustrate the effect of the density of the bed layer over this wide range for r.

At the condition $F^2 = f_3(kh_1)$, named Zone III in Figs. 2.3 and 2.4, we will have a separate stability. The growth rate results of Sect. 7 show that the Zone II growth rate increases to high values as we approach Zone III, indicating Zone III instability. However, we are at a singularity and, as shown in Sect. 11, we have zero amplitude of the interface perturbation here. Consequently, it is questionable whether we should interpret Zone III as an instability zone, as indicated by the question mark in Figs. 2.3 and 2.4. This question also seems to be of less importance in the total stability picture of the sand wave generation problem. The far more important role of the condition $F^2 = f_3(kh_1)$ is how it influences its vicinity. This is discussed in Sect. 8 below.

In Fig. 2.3, the parameter r is given values over a wide range, from 0.25 to 1.0, to illustrate the nature of the solution over this range. However, as background for seeing the value of r in actual sediment flows, Table 2.1 is established. We define the average water volume (V_w) in the bed layer by a porosity factor n, on the average volume of solids (V_s): $V_w = nV_s$. The bed layer density is thereby given by the formula: $\rho_2 = (\rho_s + n\rho_w)/(1 + n)$. Fresh water density is assumed for ρ_w, and set equal to 1.0, while the sand grains' density, ρ_s, is set equal to 2.65, 10^3 kg/m^3 being the unit of both.

Table 2.1 shows the bed layer density ρ_2 and the density ratio r over a wide range of the porosity factor n. The zero-porosity case (n = 0) is included for reference. The value of r in actual bed layer flows is likely to be below 0.3. To illustrate the stability limits for values of r in a more realistic range, Fig. 2.4 is established, covering the

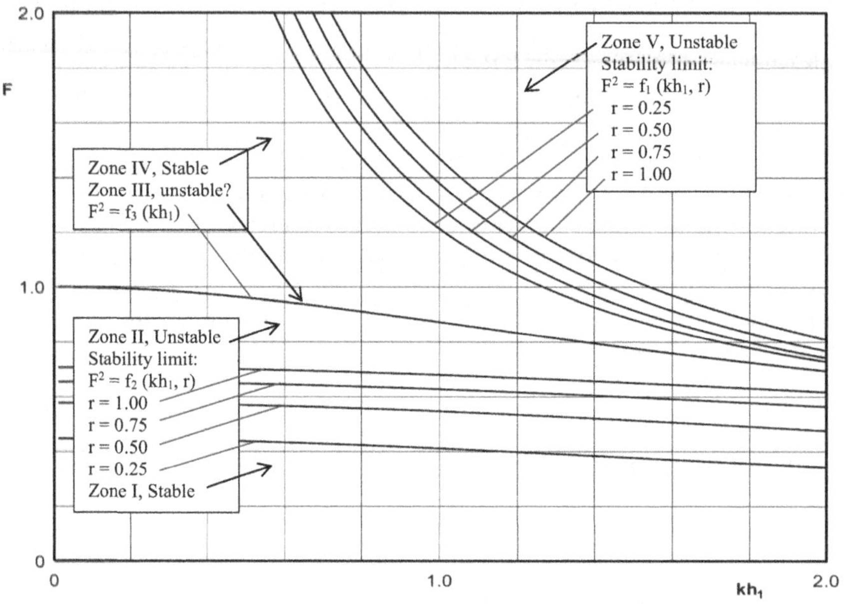

Fig. 2.3 Stability zones and stability limits for a free surface—bed layer flow

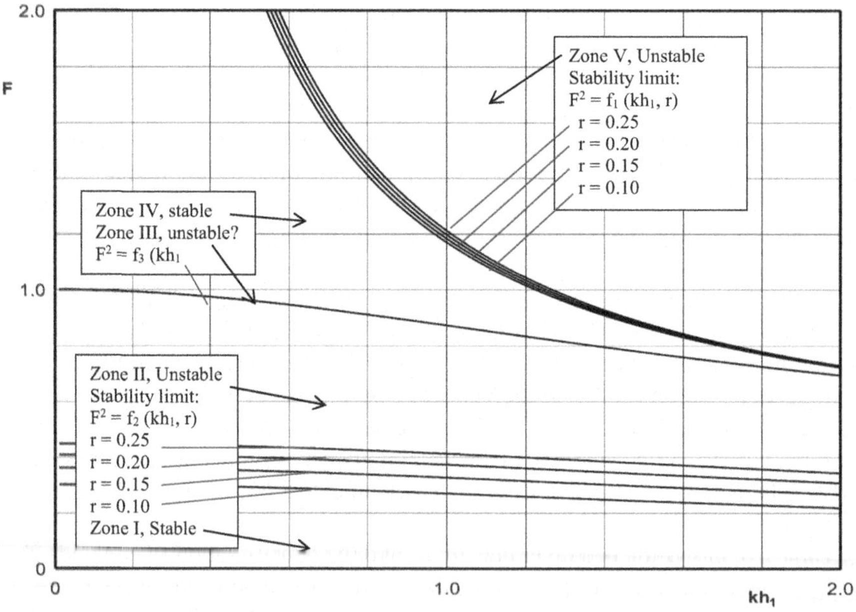

Fig. 2.4 Stability zones and stability limits for a more realistic range of r

Table 2.1 Values of ρ_2 and r for a wide range of the porosity factor n

n	0	5	7	10	12	15	20
$\rho_2/10^3$ kg/m^3	2.65	1.28	1.21	1.15	1.13	1.11	1.08
r	1.65	0.28	0.21	0.15	0.13	0.11	0.08

range $r = 0.10–0.25$. A value of $r = 0.25$ is tentatively selected for the illustrations in the following.

A main message from Fig. 2.4 is that progressive interface perturbations in flows with Froude numbers in the range: $F = 0.2–0.4$ will meet a stability limit in Zone II, where the growth rate is zero. Interface perturbations in subcritical flows with Froude number in the range: $F = 0.5–1.0$ will not meet a stability limit in Zone II, since this would require a high and rather unrealistic value of r.

We assume validity of the above stability solution for all Froude numbers in the following illustrations, while keeping the focus on the low Froude numbers.

The eigenvalue solution (2.14) may be considered also with respect to propagation rate. The second term, $O(kh_2)$, in (2.14), being always real and positive, gives a low downstream propagation rate of the unstable modes. We should also note that the first term, $O(kh_2)^{1/2}$, in (2.14), giving instability in Zones II and V, turns into a real term, i.e., a propagation rate, in the stable Zones I and IV. In both zones, the term is real and negative, expressing an upstream propagation of a stable perturbation.

7.2 *Growth Rate*

The stability solution (2.14) shows that we have a growth rate of order $(kh_2)^{1/2}$ in the unstable zones. This instability is in general a retarded instability due to the requirement $kh_2 \ll 1$. We will now take a closer look at the growth rate. For simplicity, we take the squared value of the growth rate in the stability solution (2.1). We want to study the growth rate as a function of the main flow variables F, r, and kh_1.

$$\xi_i^{\,2} = \begin{array}{c} \left\{F^2 - f_1(kh_1, r)\right\} \left\{F^2 - f_2(kh_1, r)\right\} \tanh\, kh_1 \\ \left\{F^2 - f_3(kh_1)\right\}^{-1} \left\{F^2(r+1)\right\}^{-1} kh_1\, h_2/h_1 \end{array} \tag{2.18}$$

We have introduced $kh_2 = kh_1\, h_2/h_1$. The ratio h_2/h_1 between the depths of the bed layer and the main flow is selected as 1/9, which is approximately the ratio that was observed in the verification test, which will be presented later. By solving this second order algebraic equation for F^2, and in turn for F, details omitted, we establish the diagram in Fig. 2.5. The squared growth rate $\xi_i^{\,2}$ is given parametric values varying from 0 to 10 in Zone II while covering the range 0.0 to 1.0 in Zone V. We have extended the wave number axis up to $kh_1 = 12$, with a view to the later verification test, to cover all wave lengths that might be observed.

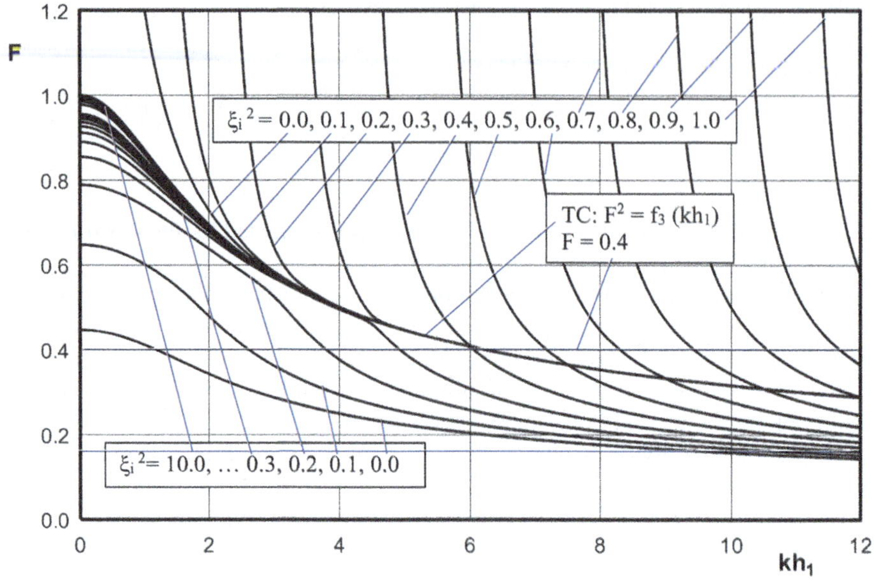

Fig. 2.5 Growth rate contours for a free surface bed layer flow, r = 0.25

We see that the stability limits $F^2 = f_1(kh_1, r)$ and $F^2 = f_2(kh_1, r)$ are included in the diagram by the parameter value $\xi_i^2 = 0$. The function $F^2 = f_3$ (kh_1) is also included. This function seems to play an important role in the stability character of this flow system. It starts from F = 1.0 with the stable Zone IV, with zero growth rate above it, and the unstable Zone II, with high growth rate just below it. The $F^2 = f_3$ (kh_1) function extends over the entire range of the diagram and will be considered in more detail in Sect. 8, where free surface effects are discussed. The name transition condition (TC) is proposed for the function $F^2 = f_3\ (kh_1) = \tanh kh_1/kh_1$.

The growth rate ξ_i for an F = 0.4 flow is determined by Eq. (2.18) and is shown in Fig. 2.6. The character of the growth rate diagram in Fig. 2.6A is an overall growth rate reducing gradually to zero at the stability limit, i.e., at $kh_1 = 1.2$, approximately.

A remarkable deviation is seen at $F^2 = f_3\ (kh_1) = \tanh kh_1/kh_1$, i.e., at TC, shown extended in Fig. 2.6B. On the Zone V side, close to the TC, the growth rate goes suddenly to zero. In a progressive development of the instability, the interface suddenly will stabilize at this stage. On the Zone II side of the TC, the growth rate goes suddenly to infinity. The infinite value at TC reflects the singularity in the growth rate function, see Eq. (2.14), and is thus considered physically non-valid.

We may note that here we have an "explosive" behavior of the interface resulting from the GKH instability. The growth rate in Zone II is seen to increase to higher

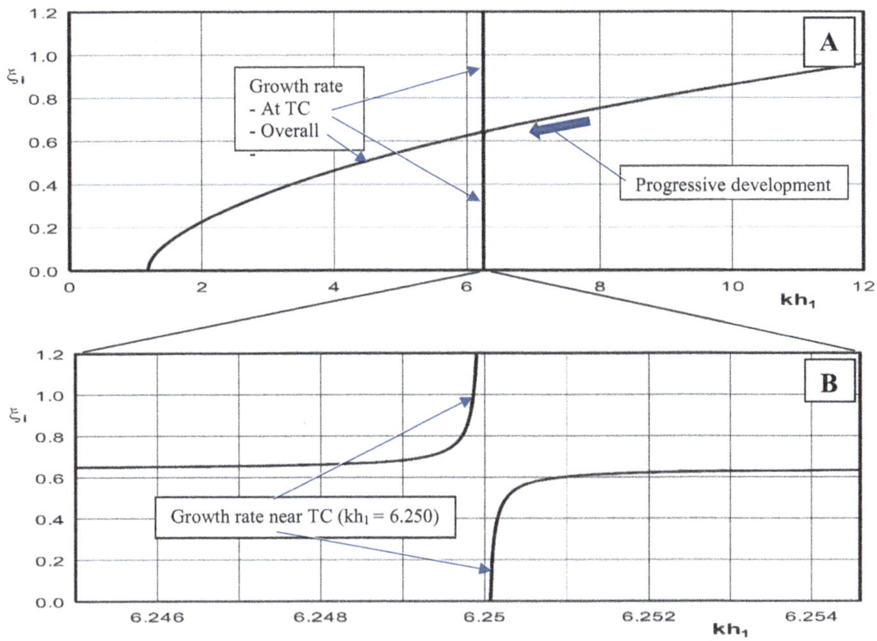

Fig. 2.6 Growth rate, F = 0.4 and r = 0.25. (**A**) Overall and at TC. (**B**) Near TC

values approaching the TC curve in the long wave range. A high growth rate approaching the TC in Zone II is therefore considered physically valid.

The growth rate changes close to the TC condition will be a key check point in the verification test. These changes represent a quite sensitive part of the growth rate function and will consequently be a good check on the theoretical solution.

8 Stability Effect of the Free Surface

8.1 Approach

The total stability effect of the free surface is studied by comparing the free surface results, already established, with the stability characteristics of a bed layer system with a fixed surface, as shown in Fig. 2.7.

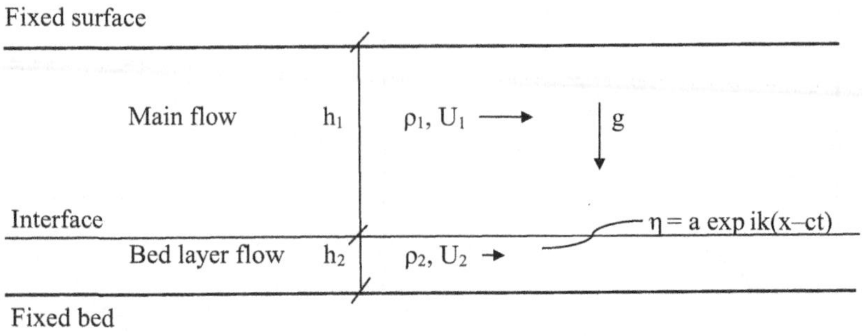

Fig. 2.7 Main flow and bed layer with fixed surface

8.2 Fixed Surface Results

The KH eigenvalue equation for a fixed surface system with finite depths is well known and may be established by applying the GKH equation (1.14) to the interface, as already done in Eq. (1.20) of Chap. 1:

$$\rho_1{}^*(c - U_1)^2 + \rho_2{}^*(c - U_2)^2 - \Delta\rho\, g/k = 0 \tag{2.19}$$

The eigenvalue solution is follows as:

$$\begin{aligned} c = c_r + ic_i = (\rho_1 * U_1 + \rho_2 * U_2)(\rho_1 * + \rho_2{}^*)^{-1} \\ +i\left\{\rho_1 * \rho_2{}^*\Delta U^2(\rho_1 * + \rho_2{}^*)^{-2} - \Delta\rho(\rho_1 * + \rho_2{}^*)^{-1} g/k\right\}^{1/2} \end{aligned} \tag{2.20}$$

The symbols are defined in Sect. 5.2. Assuming kh_2 is small, we get:

$$c_i{}^2 = \left\{\Delta U^2/\tanh\, kh_1 - (\Delta\rho/\rho_1)gh_1/kh_1\right\}(\rho_1/\rho_2)kh_2 + O\left\{(kh_2)^2\right\} \tag{2.21}$$

Equation (2.21) gives the stability limit by the requirement $c_i = 0$. Defining the same Froude number as for the free surface system, we have the stability limit, to a first approximation, given by:

$$F^2 = \Delta U^2/gh_1 = r\, \tanh\, kh_1/kh_1 \tag{2.22}$$

8.3 Comparing Stability Zones

Figure 2.8A shows the stability limits of the fixed surface system given by Eq. (2.22), while Fig. 2.8B gives the limits of the free surface system, as presented in Fig. 2.9A. We see that the free surface has a strong effect on the stability of the bed layer system by establishing two stable and two unstable zones in the F–kh_1 plane, while the bed layer system with a fixed surface has one stable and one unstable zone in this plane.

8.4 Growth Rate Comparison

Equation (2.21) for the fixed surface system is now written as follows:

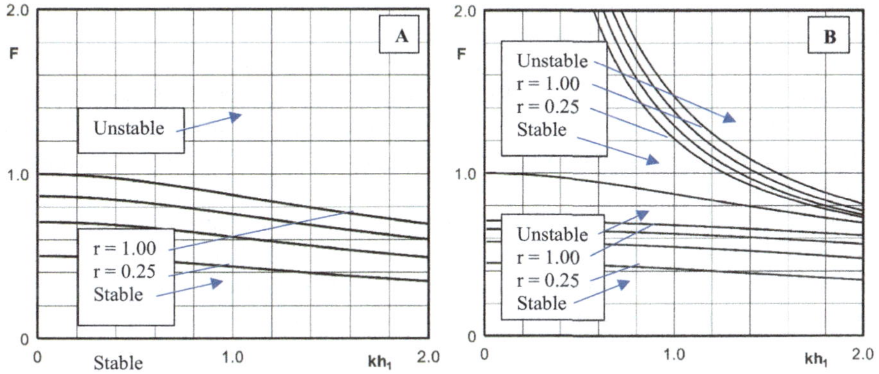

Fig. 2.8 Stability zones of flow with bed layer. (**A**) Fixed surface. (**B**) Free surface

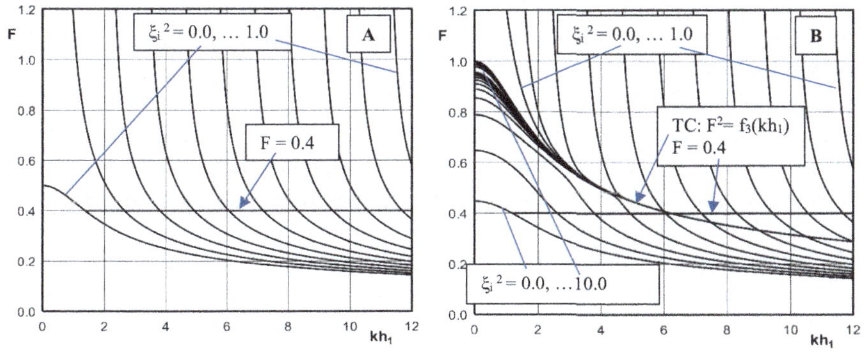

Fig. 2.9 Growth rate contours, r = 0.25. (**A**) Fixed surface. (**B**) Free surface

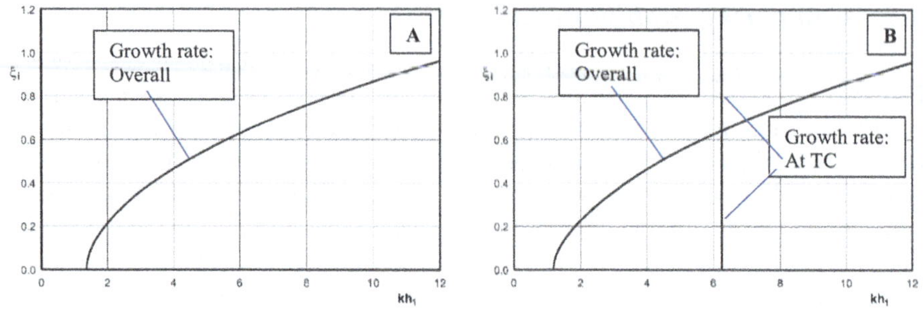

Fig. 2.10 Growth rate for F = 0.4, r = 0.25. (**A**) Fixed surface. (**B**) Free surface

$$\xi_i^{\,2} = \left(c_i^{\,2}/\Delta U^2\right) = (r+1)^{-1}\left(1/\tanh\,kh_1 - rF^{-2}/kh_1\right)kh_1\;h_2/h_1 \tag{2.23}$$

disregarding higher order terms. Solving for F and giving parametric values to ξ_i^2 as in Zone V for the free surface system, we establish the diagram as shown in Fig. 2.9A. The corresponding diagram for the free surface system, as presented in Fig. 2.10, is shown for comparison in Fig. 2.9B.

By comparing the below two figures, a clear difference is seen between the two systems. The lower stability limits for $\xi_i = 0.0$ are quite similar with slightly higher value for the fixed surface system. The F level of these curves at the long wave limit is $F^2 = r$ for the fixed surface system, while being $F^2 = r/(r + 1)$ for the free surface system.

The main difference is seen around the TC curve in the long wave range. The growth rate is zero in the stable zone IV, see Figs. 2.3 and 2.4, for the free surface system while this zone is weakly unstable for the fixed surface system, as seen in Fig. 2.9A. A remarkable and large difference is seen in Zone II for the free surface system with the growth rate going to high values as we approach the TC curve. Another remarkable observation in the comparison is the dominating role of the TC function: $F^2 = f_3\,(kh_1) = \tanh kh_1/kh_1$ for the free surface effects on the stability of long perturbations.

We now show the growth rate for an F = 0.4 flow for the two systems, given by Eq. (2.23) above for the fixed surface system, as shown in Fig. 2.10A. The corresponding figure for the free surface system is presented in Fig. 2.6 above and shown here in Fig. 2.10B.

An important observation in Figs. 2.10A and B is that the overall growth rate is nearly identical for the two systems, with a small difference for the longest waves. In both cases, we see a gradual reduction of the growth rate in a progressive development from short to longer waves.

Considering the growth rate of the free surface system, the gradual decrease as we go towards longer waves agrees with the observation of Coleman and Melville [5]. The overall growth rate, for a free surface flow at a subcritical Froude number, is then concluded to be governed by the bed layer. Except for a minor effect on the longest waves, the two overall growth rate curves are almost identical.

The free surface system has the remarkable behavior at the TC, as described for Fig. 2.6. No such effects exist for the fixed surface system. Writing the formula for the TC curve: $\Delta U^2 = gh_1 \tanh kh_1/kh_1$, it calls for a comparison with the formula for the Airy phase velocity of the gravity wave, $c^2 = gh \tanh kh/kh$, where h here is the total water depth, including the bed layer depth. For a thin bed layer, the difference between the two formulas is minor. The TC curve then expresses nearly the condition of a stationary wave on the free surface. And we see that the TC function interferes strongly with the interface stability-wise. We conclude then that we have established a stability solution explaining the accelerated growth rate observed by Jain and Kennedy [12] and later by Coleman and Melville [5]. In total the TC function: $F^2 = f_3(kh_1) = \tanh kh_1/kh_1$ expresses a strong interaction between the free surface and the interface for long perturbations.

8.5 Summary of Free Surface Results

The main results on the stability effect of the free surface are summarized as follows:

- The free surface influences the stability of the near bed flow in an alluvial channel in the long wave range. The free surface gives four stability zones in the F–kh_1 plane: two zones are stable and two are unstable. The fixed surface system has two zones: one zone is stable and the other one is unstable.
- In subcritical flow, for an instability developing progressively, the free surface causes a remarkable stability behavior for waves at the TC by suddenly stabilizing the interface just before passing the TC and by giving the interface a sudden higher instability, with an accelerated growth rate, just after passing the TC.
- The TC function: $F^2 = f_3(kh_1) = \tanh kh_1/kh_1$ has a governing role for the free surface effects on the interface stability in the long wave range.
- The overall growth rate in a free surface system is governed by the bed layer, with a minor-free surface effect on the longest waves.

9 Stability Effect of the Bed Layer

9.1 Approach

We concluded in Sect. 8 on the free surface effects that the overall growth rate is due to the bed layer. Additionally, we now look at a system where we have no effect from an upper surface, free or fixed. The purpose here is to see the stability characteristics of a system where the bed layer is governing. This system is sketched in Fig. 2.11 below. To see the effect of the bed layer on the stability of the interface in this system, we go back to the stability solution for fixed surface system as presented in Eq. (2.20) above:

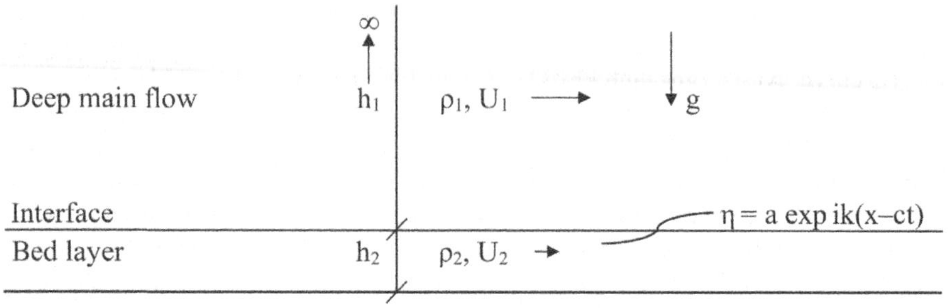

Fig. 2.11 Deep flow with bed layer

$$c_i^2 = \rho_1 * \rho_2^*(\rho_1^* + \rho_2^*)^{-2}\Delta U^2 - \Delta\rho(\rho_1^* + \rho_2^*)^{-1}g/k \tag{2.24}$$

We remove finite depth effects from a fixed—or a free—surface of the main flow, by considering it to have large main flow depth, as shown in Fig. 2.11. Assuming $kh_1 \gg 1$, the approximation $\rho_1{*} = \rho_1$ follows and we get the growth rate equation for this deep system with a bed layer of depth h_2:

$$\begin{aligned}\xi_i^2 = (c_i/\Delta U)^2 = (1+r)(\tanh\ kh_2)^{-1}[1+(1+r)/\tanh\ kh_2]^{-2} \\ - F_2^{-2}[1+(1+r)/\tanh\ kh_2]^{-1}/kh_2\end{aligned} \tag{2.25}$$

The internal bed layer Froude number F_2 is introduced by $F_2^2 = \Delta U^2/rgh_2$. The density difference r is defined in (2.7) and the tentative value 0.25 is applied as earlier.

9.2 *Deep Flow with Bed Layer. Growth Rate*

We illustrate the dimensionless growth rate, according to Eq. (2.25), to see the effect of reducing kh_2 from finite values down to small values i.e., down to waves that are increasingly long compared to the depth of the bed layer. The internal Froude number covers the range 2.0–4.0, as shown in Fig. 2.12. Figure 2.12 shows that the growth rate is strongly reduced for long perturbations. This reduction of the growth rate is caused by their proximity to the bed. We conclude that the bed layer causes a retarded instability for long perturbations on the interface. When the growth rate is reduced sufficiently, we would expect the unstable perturbation to appear as a slowly growing wave.

The influence of the internal Froude number F_2 on the growth rate, over the actual F_2 range, is seen to be minor. Figure 2.12 also shows that, for increasing wavelength

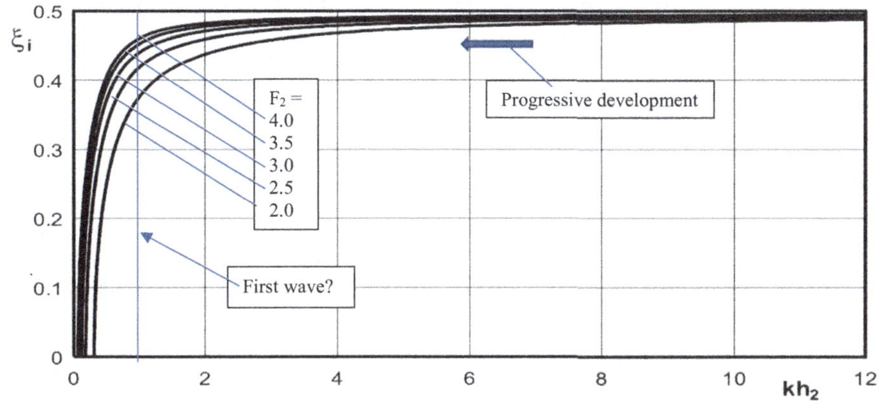

Fig. 2.12 Deep system with bed layer. Growth rate of interface wave, $r = 0.25$

of the interface perturbation, the growth rate goes to zero. Thus, the bed layer determines the stability limit for the interface perturbations.

A strong reduction of the growth rate is seen to be realized for waves longer than about $kh_2 = 1$. The first perturbations to show up as a slowly growing wave, and consequently the first sand waves to be seen, should, expectedly, be at this wavelength. For further considerations on the first waves, reference is made to Sect. 12.

From Eq. (2.25), the growth rate of short perturbations, i.e., kh_2 being large, is:

$$\xi_i = c_i/\Delta U = \left\{(1+r)(2+r)^{-2}\right\}^{1/2} = 0.497 \tag{2.26}$$

when the value 0.25 is applied on r. The stability limit, given by $c_i = 0$ in (2.25), is at the approximate wavelength (kh_2 small):

$$kh_2 = (1+r)\left[(1+r)F_2{}^2 - 1\right]^{-1} \tag{2.27}$$

or approximately $kh_2 = F_2{}^{-2}$ for the higher values of F_2.

9.3 *Summary of Bed Layer Results*

The effects of the bed layer on the KH instability of the interface are now summarized as follows:

- The bed layer strongly reduces the growth rate of the perturbation on the interface in the long wave range, from about $kh_2 = 1$. The reduced growth rate is due to the perturbation's proximity to the bed.

- From Sect. 8.5: The bed layer governs the overall growth rate of the instability, with a minor-free surface effect on the longest waves.
- The bed layer determines the stability limit of interface perturbations.

10 Propagation Rate

The eigenvalue solution (2.14) may be considered also with respect to propagation rate. The propagation rate of the unstable perturbations is given by the $O(kh_2)$ term:

$$\xi_r = \left\{ \left(F^2 - f_3(kh_1)\right)^2 + \left(1 - \tanh^2 kh_1\right)(kh_1)^{-2} \right\} \cdot \left\{ (F_2 - f_3(kh_1))^2 (r+1) \right\}^{-1} kh_1 \tanh kh_1 \, h_2/h_1 \tag{2.28}$$

Again, kh_2 is replaced by $kh_1\ h_2/h_1$. The propagation term (2.28), being always real and positive, gives a low downstream propagation rate of the unstable modes. We calculate its value for $F = 0.4$, setting $r = 0.25$ and the depth ratio $h_2/h_1 = 1/9$, as earlier. The resulting diagram is shown in Fig. 2.13A below. The growth rate diagram is shown for comparison in Fig. 2.13B. The infinite value of the propagation rate at TC is due to the singularity in the propagation rate equation and is non-physical. We note that the overall propagation rate of the unstable perturbations is gradually decreasing as the growth rate, with one main difference: At the stability

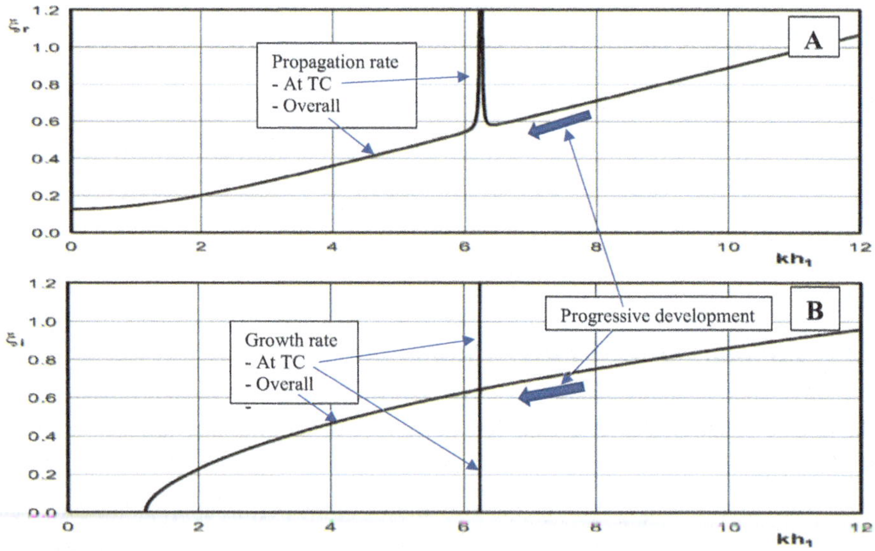

Fig. 2.13 (**A**) Propagation rate, and (**B**) Growth rate (from Figure 2.6A), of unstable perturbations for $F = 0.4$ and $r = 0.25$

limit, near $kh_1 = 1.2$, where the growth rate is zero, the propagation rate is small but still has a non-zero value. This indicates that fully developed stable dunes may propagate slowly downstream.

We should keep in mind here—from the transformation of Eq. (1.17)—that the real propagation rate is $c_r = (\xi_r + U_2/\Delta U)\,\Delta U$, where the correction term $U_2/\Delta U$ is a small number between 0 and 1. This implies that the curve in Fig. 2.13A will be lifted by the correction term to give the real propagation rate.

We should also note that the growth rate term, $O(kh_2)^{1/2}$, in (2.1), giving instability in Zones II and V, turns into a real term, i.e., a propagation rate, in the stable Zones I and IV. In both zones, the term turns real and negative, expressing an upstream propagation of a stable perturbation.

11 Transfer of the Interface Instability

11.1 Transfer to the Time Domain

Transferring the stability result to the time domain is expressed by our time domain hypothesis (TDH). Referring to the growth rate result illustrated in Fig. 2.6, the TDH states that the first unstable perturbations will develop for the shortest waves. These are short enough to develop according to the conventional KH model and will expectedly develop turbulence. Developing progressively, the perturbations will, after a while, enter the range of reduced growth rate. These perturbations will then develop as slowly growing waves.

11.2 Transfer of the Interface Instability to the Sand Bed

Transferring the stability results to the sand bed is assumed to be by the conventional scour and sediment transport mechanisms, as described in the literature on channel hydraulics, for example, the book by Raudkivi [18]. The bed erodes when the bed layer flow generates a bed shear above a critical value (Shields) on the bed surface. In a bed layer carrying sediments, we are clearly above this critical value. Increasing the bed layer velocity means increased shear on the bed, which again means increased scour. On the other hand, reduced bed layer velocity means a reduced sediment transport capacity, which again means that sediment grains will drop out of the flow and deposit on the bed. The situation for a long interface perturbation is sketched in Fig. 2.14.

The perturbation on the free surface is disregarded in the figure. Under the trough, the bed layer velocity increases, and the bed will be scoured. Under the interface wave top, the situation is opposite and sand will deposit on the bed. An increased amplitude of the interface perturbation will increase the bed layer velocity changes and thereby increase the amount of sand in this process. Thus, an increasing interface

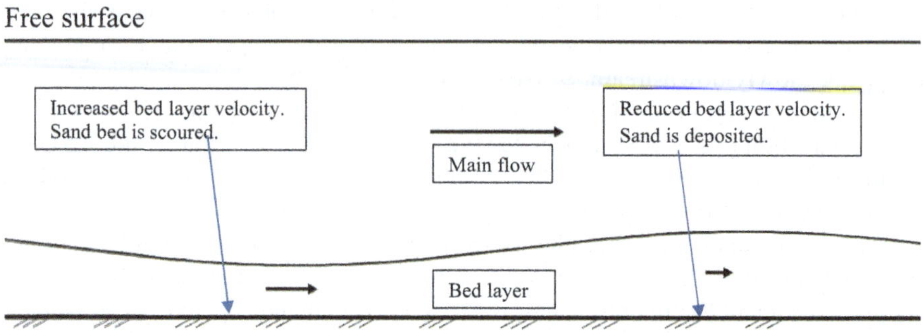

Fig. 2.14 Bed scour and deposition of sand by a long interface perturbation

amplitude, due to the instability, will increase the amount of sand in the bed layer and generate a sand wave of increasing size on the bed. The transfer mechanism thus described will clearly generate a wave form on the bed, with a certain growth rate, driven by and following the interface perturbation. This again means that the growth rate of the interface perturbation will be followed by a similar growth rate of the sand waves.

On the other hand, when the interface amplitude no longer increases, the amount of sand in this process will be constant and the growth will stop. This is the situation at the stability limit.

12 Amplitude Ratio

12.1 Amplitude Ratio in General

We now look closer at the ratio between the unstable perturbation amplitude a_2 at the interface and the associated amplitude a_1 on the free surface. We consider Eq. (2.1) where the two amplitudes are included:

$$\left\{(c-U_1)^2/\tanh kh_1 - g/k\right\}a_1 - \left\{(c-U_1)^2/\sinh kh_1\right\}a_2 = 0 \tag{2.29}$$

Introducing ζ by the definitions (2.4) and (2.5), the amplitude ratio for perturbations, satisfying $kh_2 \ll 1$, is given by the following equation:

$$\begin{aligned} a_2/a_1 &= \left\{F^2 - f_3(kh_1)(\xi-1)^{-2}\right\}F^{-2}\cosh kh_1 \\ &= \{F^2 - f_3(kh_1)(1+2\xi+\cdots)\}F^{-2}\cosh kh_1 \\ &= \{F^2 - \tanh kh_1/kh_1\}F^{-2}\cosh kh_1 + O\left\{(kh_2)^{1/2}\right\} \end{aligned} \tag{2.30}$$

since the eigenvalue ξ is $O\{(kh_2)^{1/2}\}$. We see that the amplitude ratio in general is finite and independent of r, to a first approximation. Equation (2.30) also gives that the amplitude ratio is zero at the TC: $F^2 = \tanh kh_1/kh_1$. Figure 2.15 shows that the amplitude ratio magnitude is higher or smaller than 1 in both Zones II and V.

12.2 Specific Amplitude Ratios

Two amplitude functions of special interest are outlined from (Eq. 2.30):

$$a_2/a_1 = 1: \quad F^2 = f_4(kh_1) = \sinh kh_1\{\cosh kh_1 + 1\}^{-1}/kh_1\Big\} \tag{2.31}$$

$$a_2/a_1 = -1: \quad F^2 = f_5(kh_1) = \sinh kh_1\{\cosh kh_1 - 1\}^{-1}/kh_1\Big\} \tag{2.32}$$

12.3 Amplitude Ratios Illustrated

The amplitude ratios above are illustrated in Fig. 2.15 below. The functions f_1, f_2, and f_3, defined earlier as stability limits and TC, see (2.2)–(2.4), are included for reference, smaller terms being neglected.

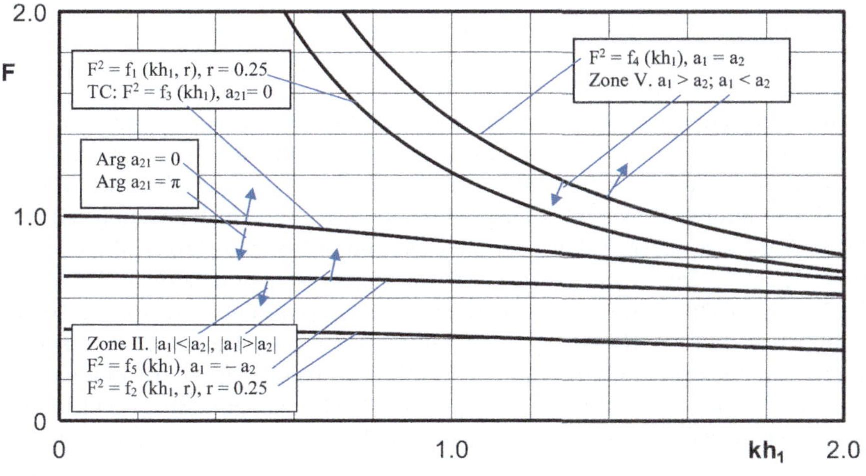

Fig. 2.15 Long wave amplitude ratio of unstable perturbations

The main results on the amplitude ratio are summarized as follows:

- The amplitude ratio of long perturbations, satisfying $kh_2 \ll 1$, has in general a finite value and is independent of r.
- The free surface and interface perturbations are antiphase below the TC and in phase above the TC.
- The amplitude of unstable interface perturbations may be larger or smaller than the amplitude at the free surface, in Zone II as well as in Zone V.
- The perturbation amplitude on the interface is zero at the TC.

When considering the amplitudes for subcritical flows, we should note that, as the perturbation develops progressively from a sudden stable state just before the TC to a sudden high instability just after the TC, the amplitude ratio goes from an in-phase to an anti-phase state.

13 Validity Ranges

The theoretical results established in the preceding are subject to validity limitations. We have focused on subcritical flow and established a stability solution in the long wave range based on the requirement $kh_2 \ll 1$. This requirement may be satisfied also by flows with higher Froude numbers. This will include determination of h_1, h_2 and thereby the Froude number, which now will be different from the hydraulic Froude number. The Froude number range, where the long wave results are valid, is not studied further here but should be clarified in future work.

The range of free surface effects on the bed layer stability is another question of high interest. The present theory does not give the limit explicitly. However, Fig. 2.5 shows clearly a weakening free surface effect as we go towards shorter waves along the TC curve. Future laboratory testing will give the answer here.

A third question of high interest is the wave number range where the long wave results are valid, i.e., what is the maximum value of kh_1 for application of the established long wave results? A basic premise in developing these stability results is the requirement $kh_2 \ll 1$ or $2\pi h_2 \ll \lambda$, λ being the wavelength. In physical terms, this is a requirement saying that the perturbation wave shall be long compared to the depth of the bed layer. Taking the physical thinking a step further, the condition $kh_2 = 1$, or $\lambda = 2\pi h_2$, expresses that the wavelength is 2π or 6.28 times the depth of the bed layer. It seems to be a reasonable thought that the growth rate of this perturbation will be affected by its proximity to the bed. The established stability results may thus hopefully be valid up to $kh_2 = 1$. This criterion, if valid, will then determine the first perturbation to develop as a slowly growing wave. By the transfer mechanism to the sand bed, this criterion will then determine the length of the first sand waves to be observed. This will be another important check point in the verification test. The first wave criterion, if/when validated, may then be turned around: the length of the first sand waves determines the effective depth of the bed layer.

14 Theoretical Work Summary

The theoretical solution for the stability of a channel flow with a bed layer is developed. The established theoretical results are based on stated hypotheses, listed premises, and a specific discretized theoretical flow model for the study. Stability analysis of the flow system has been performed by applying the General Kelvin-Helmholtz stability model. The analysis has resulted in various stability characteristics:

- Four stability zones in the Froude number-wave number plane
- Growth rate characteristics for a retarded long wave instability
- Influence of the free surface on the interface stability
- Influence of the bed layer on the interface stability
- A mechanism transferring the instability from the interface to the sand bed
- Propagation characteristics for the unstable perturbations
- Perturbation amplitude relations between the free surface and the interface

Some of the theoretical results are in good agreement with results in earlier work. Final conclusions will be stated based on the observations in a verification test, together with results from earlier work. The verification test performed as part of the present study is described in the following.

15 Verification Test

15.1 Test Objective and Testing Philosophy

The objective of the test is to observe visually the generation of sand waves in subcritical channel flow and compare the observations with the theoretical stability results.

Running the tests on a high level is considered saisfactory to meet this objective. This means that no detail measurements will be made while concentrating on visual observations of the main stability characteristics. The number of test parameters will be kept to a minimum, while their value will be determined approximately.

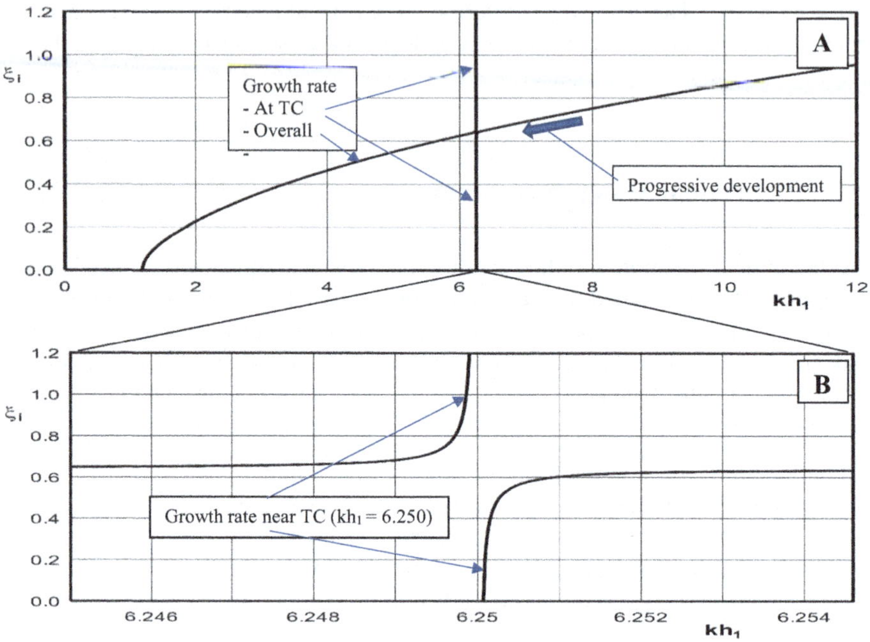

Fig. 2.16 Growth rate diagram for interface perturbations in a bed layer channel flow. (**A**) Overall growth rate. (**B**) Growth rate near TC

15.2 Theoretical Reference and Theory Check

The main theoretical reference will be the growth rate diagram established in Fig. 2.6, shown again in Fig. 2.16.

The following theory check points for the visual observations were established:

1. The thickness of the sediment carrying bed layer as set up by the flow
2. The length of the first sand waves
3. Zone V progressive generation of sand waves, i.e., longer sand waves develop progressively from the first ones
4. The stability of the near bed flow just before the TC
5. The increased instability of the near bed flow just after passing the TC
6. Zone II progressive generation of sand waves
7. The final stage of the generated sand waves
8. The overall growth rate
9. Transfer of instability to the bed
10. Propagation rate check
11. Amplitude check for subcritical flow

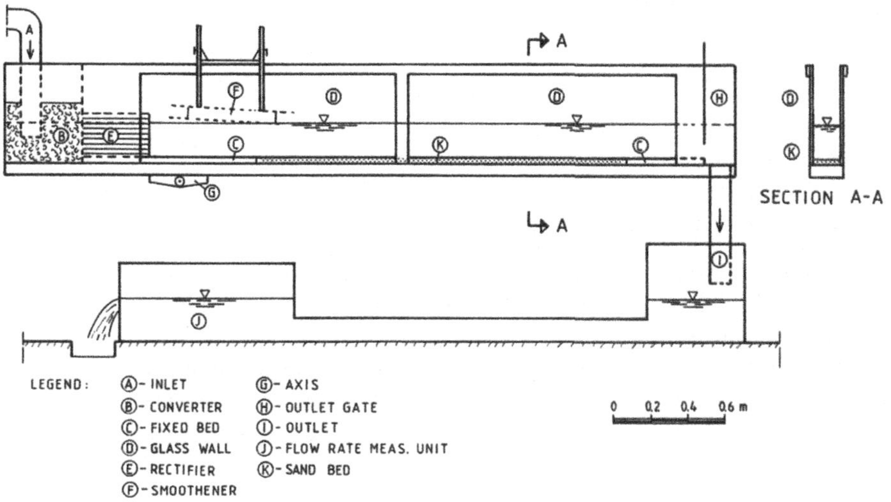

Fig. 2.17 Test setup in laboratory flume

15.3 Testing Flume

The test setup is shown in Fig. 2.17 below. The flume was 4 m long, 60 cm deep, and 20 cm wide.

To have a check on 2D theoretical results, it will be important to have 2D flow conditions in the test flume. For this purpose, the flume was equipped with flow rectification and smoothening devices upstream of the test section. The test section with sand was 2 m long and was made smooth and even before each test.

Initial calibration tests showed quite uniform flow conditions with good visibility through the glass walls and through the flowing water. A white board with a 5 cm × 5 cm grid was mounted on the rear glass wall. The grid served well for visual determination of approximate height and length of the sand waves.

15.4 Sand for Testing

A test section along the bed of the flume was filled with sand, as shown in Fig. 2.17. Sieve testing of the sand was performed. The sand was found to be nearly one-graded with a D_{50} of 0.20 mm. Sieve testing diagrams are included in Ref. [15].

15.5 Testing Method

The tests were performed working with the following four test parameters:

- The total water depth
- The hydraulic Froude number
- The depth of the bed layer
- The length of the sand waves

Thus, the growth rate of the sand waves is not determined in these tests. The total water depth and the hydraulic Froude number were set by calibrating the flume prior to the test. The bed layer depth and the wavelength of the sand waves were determined by the visual observations. Video photographic equipment was used to produce the visual documentation of the generation of the sand waves. The visual observations followed the sand wave generation process from test start to test end.

The test runs were performed by two main activities:

- Starting up the water flow in the flume to establish the flow with the specified Froude number
- Thereafter keeping the flow constant while performing visual observation of the sand wave generation process in the flume

This testing method is clearly on a high level, in accordance with the testing philosophy. All the above test parameters are considered reasonably well determined, both the preset and the observed ones, while the test method naturally will give approximate values. However, the test method is considered valid for serving the test objective.

15.6 Test Program

The test program was to run a limited number of runs with F = 0.4 (base case) to establish a good basis of video recordings. Additionally, we decided to run a specific test at F = 0.6 for the subcritical amplitude check.

15.7 Base Case Test Observations

15.7.1 Initial Stage

For each base case run an initial stage lasted for a short while where a bed layer carrying sediments, without sand waves, was observed. The bed layer depth was visually assessed to be about 1 cm in each of the base case runs.

Fig. 2.18 F = 0.4. First sand waves

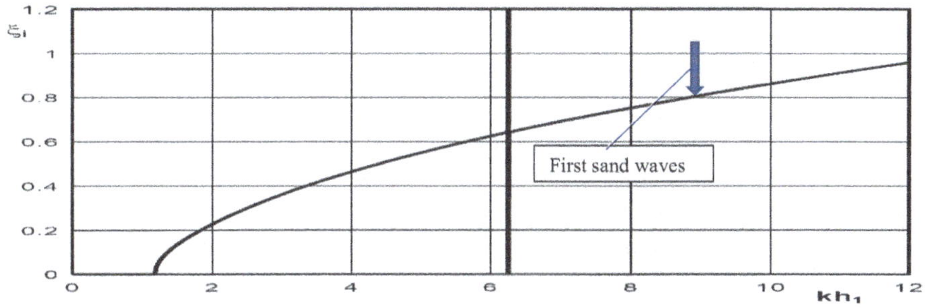

Fig. 2.19 Approximate wavelength of the observed first sand waves

During this initial stage, no bed waves were generated. The duration of this stage was not measured. However, we are talking about less than a minute.

15.7.2 First Sand Waves

The sand wave generation process was initiated by occurrence of the first waves terminating the initial stage, in all base case runs. The first sand waves appeared at the same moment along the entire test section. They were quite regularly ordered, i.e., having about the same height and the same wavelength. The transverse configuration of the waves was nearly 2D. No interaction with the free surface was observed at this stage. First sand waves in a base case test are shown in Fig. 2.18. The photo is taken out from the video shortly after the first sand waves occurred. The blue arrow gives the flow direction, while the 5 cm × 5 cm grid is seen in the background. Three sand waves are seen to have a total length of about 20 cm, giving an average wavelength of approximately 6.7 cm. The approximate wavelength of the first sand waves observed is indicated by the blue arrow in the response diagram of Fig. 2.19.

15.7.3 Progressive Development in Zone V

The first sand waves developed progressively, i.e., by increasing gradually their height and length. The progressive development continued, and after a good while (minutes), the neighborhood of the TC was entered.

15.7.4 Observations at TC

The sand waves stopped increasing their amplitude and length, while the near bed flow appeared quite stable. This situation was clearly observed visually but is not well documented by any available photo.

The stable situation lasted for a short while whereafter the near bed flow turned into an excited mode. Bed sand was lifted violently up at certain locations while the excited flow spread it into suspension. The picture in Fig. 2.20A is taken locally early in the excited flow stage, with a blue vertical arrow indicating a concentrated local flow upwards, a kind of "lift-up" of the bed sediments, which was not seen during the earlier stages of the sand wave generation. We should keep in mind here that the perturbation at this state suddenly turns excited and at the same time suddenly shifts to antiphase, see Sects. 7 and 11.

The excited flow appeared quite 3D and the sand bed surface turned quite irregular during the excited phase. A local part of the resulting irregular bed is shown in Fig. 2.20B. The two stages are indicated in the response diagram, Fig. 2.21, keeping in mind that the infinite limit and the higher parts of the theoretical curve are not considered physically valid.

The length of the sand waves at this stage was not assessed from the video due to the 3D and highly excited near bed flow and the irregular bed, as described above. The important observation here is that this increased (or accelerated) growth rate in fact took place, as predicted by the present 2D theory.

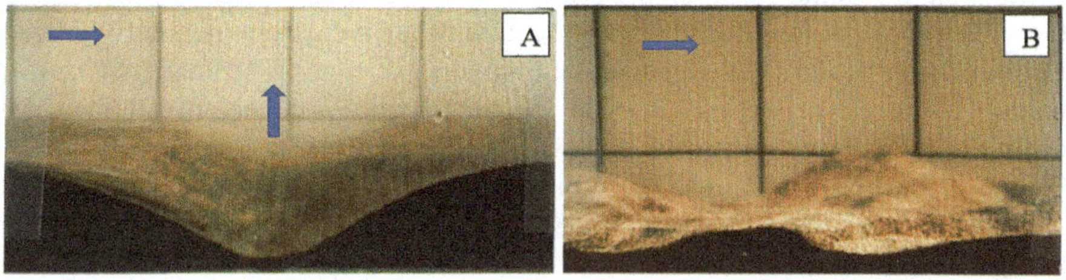

Fig. 2.20 F = 0.4. (**A**) Excited near bed flow with "lift-up." (**B**) Irregular bed

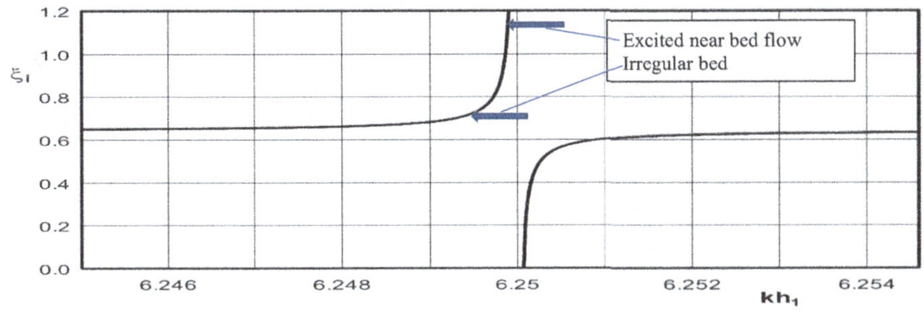

Fig. 2.21 Approximate stages of excited flow and resulting irregular bed

15.7.5 Observations in Zone II

The excited near bed flow lasted for a while (likely less than a minute), then calming and turning gradually into a sand wave generation mode again.

The sand waves now were longer than before and continued developing increasing height and length, confirming an ongoing progressive instability. The sand waves at this stage, according to the theory, are generated by the Zone II instability and had the typical shape of sand dunes, propagating downstream. Their shape was again nearly 2D and quite regularly ordered along the test section.

The situation shown in Fig. 2.22 is a good while (minutes) after the TC stage and a bit prior to the final stage. The approximate wavelength at the development stage in Fig. 2.22 is indicated in Fig. 2.23.

The orange curve in the photo is introduced artificially, indicating an interface perturbation of the same length and amplitude as, and in antiphase with, the wave on the free surface, assuming its validity for this non-horizontal bed. The curve is introduced for considering the transfer mechanism, see Sect. 15.7.7 below.

The sand waves propagated downstream during the entire Zone II development.

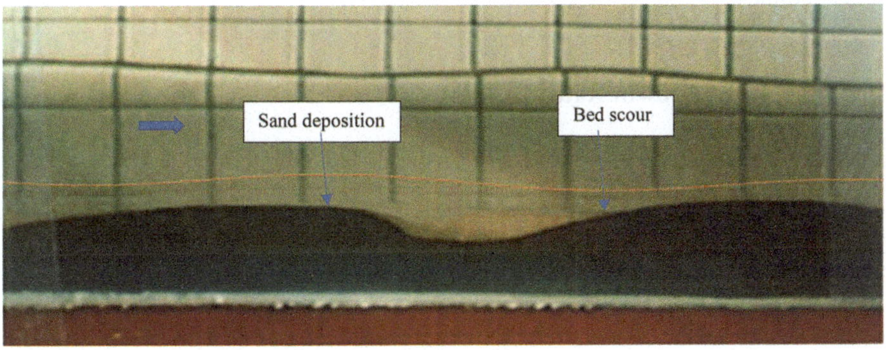

Fig. 2.22 Developed dunes in Zone II. Bed scour and deposition of sand

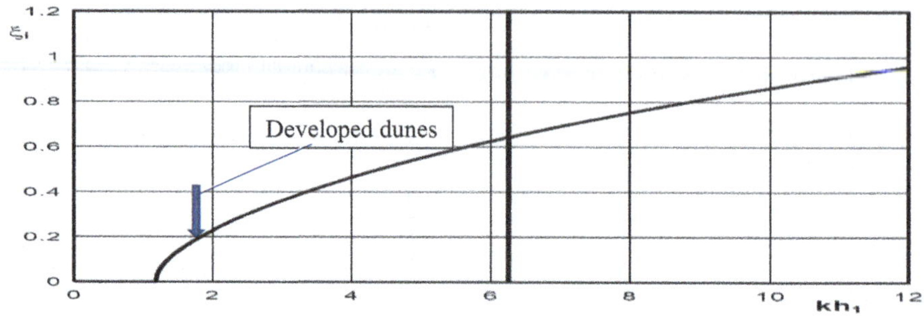

Fig. 2.23 Approximate wavelength of the dunes in Fig. 2.22

15.7.6 Final Dunes

The Zone II progressive development of dunes continued until they finally stopped growing. Two observations are noted:

- The final sand dunes were stable, without changing amplitude or wavelength.
- The final dune length was about 35 cm.

According to the theory, the final dunes will be at the stage where the $F = 0.4$ level meets the Zone II stability limit, see Fig. 2.5.

15.7.7 Transfer Mechanism

The theoretical locations of sand release and sand settlement under an antiphase interface wave, by the transfer mechanism described in Sect. 10, are shown in Fig. 2.22. Two observations are noted:

- Release and settlement of sand grains were observed, both in the video and by direct visual observation during the test, at these locations.
- The sand bed deposition location is confirmed in the photo by the shape of the sand dunes.

15.7.8 Overall Growth Rate

The main observations on the overall growth rate from the first sand waves until the final dunes in our base case tests were the following:

- The first sand waves had a certain growth rate.
- The final dunes were stable, i.e., with zero growth rate.

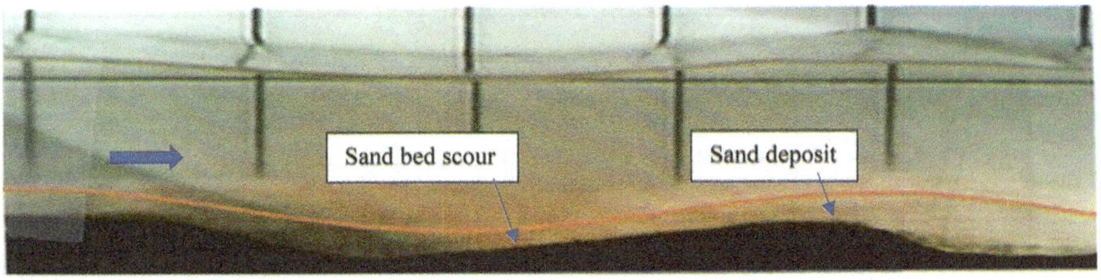

Fig. 2.24 Sand wave and free surface wave in phase, Zone V, F = 0.6

15.7.9 Propagation Rate Check

Two main observations were made on the propagation rate of the sand waves:

- In the base case tests, the sand waves propagated downstream in Zone V, prior to the TC crossing, as well as in Zone II, after the TC crossing.
- In the amplitude check test, see Sect. 15.8, the Zone V sand waves propagated downstream.

15.8 Amplitude Check

A separate test was run to check the Zone V amplitude function for subcritical flow. Figure 2.24 shows the sand waves and the free surface at a stage prior to the TC in Zone V, for an F = 0.6 (approximately) test. We see that the free surface is in phase with the sand wave in this subcritical flow.

Figure 2.25 gives the approximate wavelength in the extended amplitude diagram. We conclude that this in-phase amplitude relationship agrees with the present theory.

An interface wave with slightly larger amplitude than the surface wave, in accordance with the $a_1 < a_2$ domain in Fig. 2.25, is introduced artificially in the photo. The locations where sand bed is scoured and where sand is deposited on the bed are shown. The shape of the sand wave below is seen to be well in line with this. Again, the interface instability seems to be transferred to the bed in accordance with the transfer mechanism, as described in the preceding.

The sand waves propagated downstream in the amplitude check test.

15.9 Observation Summary and Evaluation

The observations are now summarized and evaluated following the order of the above check points as stated in Sect. 15.2.

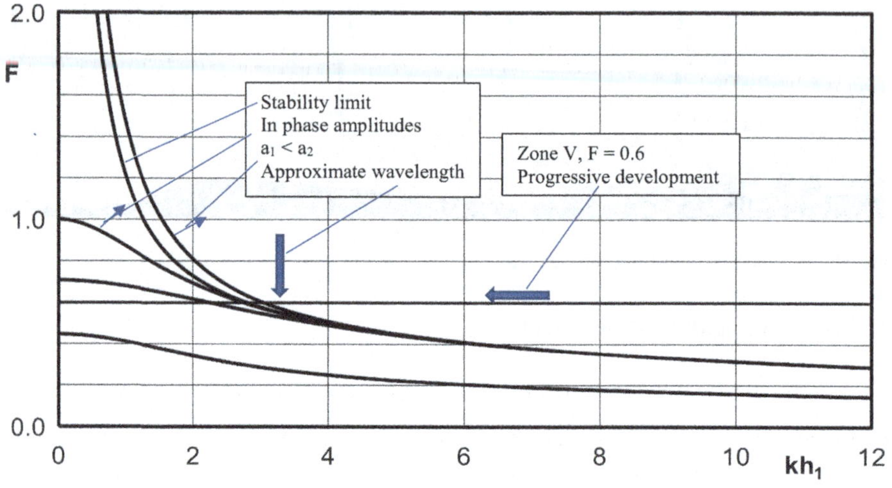

Fig. 2.25 Approximate wavelength and Froude number for Fig. 2.24

1. The depth of the sediment carrying bed layer was about 1 cm for the base case tests, giving a depth ratio $h_2/h_1 = 1/9$. The visually observed bed layer depth is of course approximate.
2. An average length of the first waves observed in the test was 6.7 cm, while the first wave criterion, $kh_2 = 1$, gives 6.28 cm, both figures being approximate. These two values agree surprisingly well. The apparent agreement may be accidental. However, we propose the first wave criterion tentatively established. Future work should clarify and may confirm the criterion.
3. A progressive development was observed for all sand waves in Zone V.
4. The observed Zone V sudden stability is not well documented by the present study. This is a point to be confirmed by future work.
5. The Zone II suddenly excited near bed flow close to the TC was very clearly observed in agreement with the theoretical stability solution. The 3D character of the excited flow is naturally not described by the 2D theoretical model. However, the higher instability is described by the 2D model. The Zone II increased instability also agrees well with the conclusions on accelerated growth by Jain and Kennedy [12] and later by Coleman and Melville [5]. They concluded that the accelerated growth rate takes place near the condition $F^2 = \tanh kh/kh$, which also means, as already stated, it takes place near the condition $F^2 = \tanh kh_1/kh_1$, which is the present theoretical requirement. The excited near bed flow observations near the TC in Zone II are seen as providing a good support for the theoretical model and the derived stability results.

We also note that the observation by Coleman and Melville that the period of accelerated growth was flanked by periods of slower development is in good agreement with the growth rate curve in Fig. 2.23.

6. The progressive character of the sand wave generation continued in Zone II until the final stability limit. This means that the progressive development of Zone V continued beyond the Zone V stability limit, it continued across the narrow stable Zone IV, and it continued as across the TC. Thus, the instability develops progressively over the entire wavelength range from the first sand waves in Zone V until the final and stable dunes in Zone II. The progressive development of sand waves is also described by the work by Jain and Kennedy [12] and later by Coleman and Melville [5].
7. The final dunes were stable, i.e., having zero growth rate with no further growth of amplitude or wavelength, thus confirming the Zone II stability limit. The final wavelength of the dunes is theoretically for $kh_1 = 1.2$, as seen in Fig. 2.5. With $h_1 = 9$ cm and $F = 0.4$, this gives a final wavelength of 47 cm while the observation gave about 35 cm. The final wavelength depends on both r and F and will consequently be influenced by the uncertainties we have on r and F. Additionally we may consider that the large dunes may reduce the Froude number of the flow. This will result in a reduced length of the final dunes. This is another point to be checked in future work.
8. The overall growth rate is seen as largely in line with the growth rate described by the theory. Even if this is as a rather rough visual measure on the overall growth rate, it may be said to agree largely with the theoretical diagram in Fig. 2.23. This diagram may also be said to agree well with Coleman and Melville's [5] work, where they reported a gradual reduction of the growth rate from the initial wavelets to the final dunes.
9. Observations on the developed dunes in Zone II and on the Zone V bed forms in the amplitude check test both give good indications that the transfer mechanism works as described in the theory.
10. The observations on the downstream propagation of the sand waves, driven by the interface perturbations, in Zone V as well as in Zone II, are in line with the theoretical solution in both zones.
11. With the uncertainties on the Froude number and the wave number in the amplitude check test, the test is still concluded to confirm the in-phase relation between the interface and the surface waves for this flow, which is well down in the subcritical regime.

16 Study Conclusions

Acknowledging the missing documentation of the sudden stability just before the TC, the above-evaluated observations are seen as giving a quite good support to the other main characteristics of the theoretical model and the derived theoretical results. The conclusions from our study on sand wave generation in subcritical flow are thus summarized as follows:

- Subcritical flow in alluvial channels, at low Froude numbers, generate sand waves by a hydrodynamic instability of the bed layer interface. The instability is described by the General Kelvin-Helmholtz stability model.
- Sand waves are generated by a retarded instability of interface perturbations that are long compared to the depth of the bed layer. Their growth rate is reduced by their proximity to the sand bed.
- The instability is transferred from the interface to the sand bed by a scour and deposition mechanism. The bed layer's scour and deposition of sand is governed by the local bed layer velocity, which again is governed by the long and slowly growing unstable perturbation at the interface.
- The interface instability develops progressively from the first short perturbations (ripples) until the final stability limit, determined by the bed layer, where stable and fully developed dunes is the result on the sand bed.
- The free surface interacts with the long perturbations at the bed layer interface, by:
  - Determining four stability zones in the Froude number—wave number plane
  - A sudden stability and instability effect on interface perturbations at the transition condition
- The bed layer has a governing effect on the interface instability generating sand waves, by:
  - Determining the overall growth rate of the instability
  - Reducing the growth rate of the instability in the long wave range
  - Determining the stability limit for the long perturbations
- The overall growth rate of the unstable perturbations, being governed by the sediment carrying bed layer, decreases gradually from a finite value for the first waves until zero at the stability limit.
- A criterion determining the length of the first sand waves is tentatively established.
- The sand waves, in Zone II as well as in Zone V, propagate downstream.
- Sand wave and free surface amplitudes for subcritical flow, at a pre-TC stage, are in phase.

17 Final Remarks

By these conclusions, the Main Study Hypothesis (MSH) is considered confirmed. The time domain hypothesis (TDH), and thereby the observation by professor Raudkivi [16], is also confirmed. This was already done by the works of Jain and Kennedy [12] and later by Coleman and Melville [5]. The MSH conclusion is established by an approach that deviates rather basically from the approaches taken in earlier studies, thus inviting some remarks.

Firstly, according to the present results, the driving instability generating sand waves is developed at the bed layer interface. The instability is transferred to the sand bed, by a local scour and deposition mechanism in the bed layer.

Secondly, the stability model applies discrete fluid layers with constant velocity of each layer. The vorticity is vertically integrated and included at the bed layer interface and at the bed surface. One may think that this representation of the velocity field is overly simplified, see, for example, Yih, p. 241. However, the above results of the present study seem to prove otherwise.

Thirdly, the widely used Brooke Benjamin [4] maximum growth rate criterion is clearly not applicable to the interface instability generating sand waves on a flat bed, according to the present results. On the contrary, sand waves are generated by a progressive and retarded instability with a strongly reduced growth rate. Additionally, the final and fully developed dunes are present at a state of interface stability, i.e., at zero growth rate. Thus, fully developed sand dunes on the bed are the resulting stable bed features developed by an earlier progressive and retarded instability at the bed layer interface.

As a point number four, the progressive instability is a challenging stability concept, calling for future interest.

As a point number five, the theoretical study is strictly 2D. While natural sand waves, in particular ripples, often have a remarkable 2D character, sand waves also develop 3D characteristics subject to a 3D flow environment. It should be noted, however, that the 2D theory, in combination with appropriate 2D testing, in fact opens for the confirmation and understanding of the sand wave generation mechanism.

Finally, it should be noted that the presented stability results describe an initial development only, as is the case for all results of linear stability analysis. The further development of the instability requires non-linear analysis or non-linear computer models, as already pointed out.

18 Further Work

The present study calls for a lot more work. First, of all, running the same 2D tests as the present ones, including video recording, for the purpose of verification of the present results, is warmly recommended, for two good reasons:

- Observing the remarkable agreement between the sand wave development and the theoretical results
- Observing its firm support of the GKH stability model

The following points for extending the solution, by theoretical work and by laboratory testing, should also be considered:

- Extending the present work to higher Froude numbers, i.e., further subcritical as well as supercritical flow cases.

- Checking a possible reduction of the Froude number by the large dunes. The consequence may be a shorter length of the final and stable dunes.
- Checking a possible reduction of the Froude number by the anti-dunes. By this, anti-dunes develop into Zone IV, as stable sand waves propagating upstream.
- Checking ripples coexisting with dunes. One type of test may be by stopping the flow in a progressive development of dunes whereafter the flow is restarted. While the developed dunes remain, new ripples may be generated on top of the dunes.
- The conditions when the progressive development of sand waves, in an ongoing flow, starts over again, with generation of new first sand waves.
- Checking how the described mechanism works for a different grain size of the bed sand.

The results presented above will, hopefully, give a contribution to understanding the physical mechanism generating sand waves in alluvial channels and other environments. This will then be useful in further scientific work and in developing the tools necessary for engineering work on sand waves in all types of environments.

References

1. Ali, Sk Zeeshan, and S. Dey. 2021. Linear stability of dunes and anti-dunes. *Physics of Fluids* 33: 094109.
2. ASCE Committee on Sedimentation. Task Force on Bed Forms in Alluvial Channels. 1966. Nomenclature for bed forms in alluvial channels. *Journal of Hydraulic Engineering* 92 (3): 51–64.
3. ASCE Task Committee on Flow and Transport over Dunes. 2002. Flow and transport over dunes. *Journal of Hydraulic Engineering* 128 (8): 726–728.
4. Brooke Benjamin, T. 1957. Wave formation in laminar flow down an inclined plate. *Journal of Fluid Mechanics* 2: 554–574.
5. Coleman, S.E., and B. Melville. 1994. Bed-form development. *Journal of Hydraulic Engineering* 120 (5): 544–560.
6. Coleman, S.E., and B. Melville. 1996. Initiation of bedforms on a flat sand bed. *Journal of Hydraulic Engineering* 122 (6): 301–310.
7. Colombini, M. 2004. Revisiting the linear theory of sand dune formation. *Journal of Fluid Mechanics* 502: 1–16.
8. De Jong, B. 1989. Bed waves generated by internal waves in alluvial channels. *Journal of Hydraulic Engineering* 115: 801–817.
9. Engelund, F. 1970. Instability of erodible beds. *Journal of Fluid Mechanics* 42 (2): 225–244.
10. Engelund, F., and J. Fredsøe. 1982. Sediment ripples and dunes. *Annual Review of Fluid Mechanichs* 14 (1): 13–37.
11. Fredsøe, J. 1974. On the development of dunes in erodible channels. *Journal of Fluid Mechanics* 64 (1): 1–16.
12. Jain, S.C., and J.F. Kennedy. 1971. The growth of sand waves. In *Proceedings of International Symposium on Stochastic Hydraulics*, 449–471. Pittsburgh: Pittsburgh University Press.
13. Kennedy, J.F. 1963. The mechanics of dunes and anti-dunes in erodible-bed channels. *Journal of Fluid Mechanics* 16 (4): 521–544.
14. Liu, H.-K. 1957. Mechanics of sediment-ripple formation. *Journal of the Hydraulics Division* 83 (2): 1–23.

15. Moshagen, H. 1984. Sand waves in free surface flow. A study with emphasis on hydrodynamic aspects. Doctoral thesis, NTNU, Trondheim, Norway. https://ntnuopen.ntnu.no/ntnu-xmlui/handle/11250/2622275.
16. Raudkivi, A.J. 1966. Bed forms in alluvial channels. *Journal of Fluid Mechanics* 26 (3): 507–514.
17. Raudkivi, A.J. 1983. Thoughts on ripples and dunes. *Journal of Hydraulic Research* 21 (4): 315–321.
18. Raudkivi, A.J. 1990. *Loose Boundary Hydraulics*. Oxford: Pergamon Press.
19. Reynolds, A.J. 1965. Waves on the erodible bed of an open channel. *Journal of Fluid Mechanics* 22 (1): 113–133.
20. Richards, K.J. 1980. The formation of ripples and dunes on an erodible bed. *Journal of Fluid Mechanics* 99 (3): 597–618.
21. Sumer, M., and B. Bakioglu. 1984. On the formation of ripples on an erodible bed. *Journal of Fluid Mechanics* 144: 177–190.
22. Venditti, J.G., M. Church, and S.J. Bennett. 2006. On interfacial instability as a cause of transverse subcritical bed forms. *Water Resources Research* 42: W07423.

Chapter 3
Generation of Water Waves by Wind

1 Background and Objective

1.1 Study Background

1.1.1 Earlier Work

The mechanism by which wind generates waves on the water surface is a geophysical problem that has attracted interest over many decades. Early papers on the subject date back to the second half of the nineteenth and the first half of the twentieth century.

The Kelvin-Helmholtz instability was applied to the problem of wave generation by wind by Lamb [2, pp. 374 and 462]. With gravity as the only stabilizing force, he established that all wavelengths shorter than a certain value would be unstable. And he states that if there were no modifying circumstances, the slightest breath of wind would ruffle the surface of water. However, by including the stabilizing effect of surface tension together with gravity on the water surface, he determined the minimum phase velocity of surface waves. A corresponding wind velocity necessary for wave generation was thereby determined to be about 650 cm/s. However, wind velocity in the range 1–2 m/s is observed to generate small waves, see, for example, Bye and Babanin [3]. Thus, application of the conventional Kelvin-Helmholtz stability model does not explain the initial waves on the water surface generated by the wind.

A considerable effort was made during the 1950s and 1960s. Chia-Shun Yih writes a summary of the status by 1980 in his book on Stratified Flows [1]. Yih refers to Ursell, 1956 [7], where he stated that the present state of knowledge was profoundly unsatisfactory. Yih summarizes the works by Jeffreys (1924–1925), Eckhart (1953), Phillips (1957), Lighthill (1962), Longuet-Higgins (1962), and Miles (1957–1962). Miles' work was given support by the results from field measurements reported by Longuet-Higgins, Cartwright, and Smith (1961).

H. Moshagen, *The General Kelvin-Helmholtz Stability Model*,
https://doi.org/10.1007/978-3-031-61984-7_3

Miles stated his objective in the Introduction to his first and basic paper [5], being the prediction of the energy transfer from a prescribed two-dimensional parallel shear flow in an inviscid incompressible fluid – i. e. a wind field – to a surface wave of prescribed wavelength and wave speed.

Some of the key points of Miles' theory are now pointed out:

- A two-dimensional turbulent wind field is assumed with a mean shear flow U(y). The velocity and pressure fields are given a small disturbance allowing linearization of the equations, while the turbulence fluctuations are neglected in the perturbation equations. The inviscid Orr-Sommerfeld equation is the result of these assumptions and is applied for his following stability analysis.
- The water body is assumed to be inviscid, irrotational, and incompressible with a small slope of the perturbation waves on the water surface.
- The aerodynamic pressure is assumed to be:

$$p_a = (\alpha + i\beta)\rho_a U_1^2 k \eta \tag{3.1}$$

  where ρ_a the density of the air, U_1 is a reference wind velocity, k is the wave number of the perturbation, and η is the displacement of the water surface. The part β of the pressure, being in phase with the wave slope, is the pressure part of importance. Miles states that the β assumption rules out the Kelvin-Helmholtz mechanism of wave generation a priori.
- Miles found that the rate at which energy is transferred to a wave of speed c is proportional to the curvature of the wind velocity profile—U'' at the elevation where $U = c$.

In a later publication [6], Miles proposed a generalized Kelvin-Helmholtz model where the wind field had a vertical velocity profile U(y). Miles applied his generalized K-H model to the problem of wind generation of waves on a viscous liquid (oil) as well as to the water surface. Miles concluded that his generalized K-H model appeared adequate to explain the instability of the interface between a viscous liquid and an approximately parallel shear flow in the air above. It appeared, on the other hand, unlikely that the K-H mechanism, according to his generalized model, could be significant for liquids of low viscosity as wind over water.

Bye and Babanin [3], in their article "Wave Generation by Wind" in Encyclopedia of Ocean Sciences, 2009, state that the first wave to be generated as the wind speed increases is likely to be the wave of minimum phase speed, propagating at an angle to the wind direction. Laboratory observations indicate that at slightly higher wind speeds, wave growth results from a shear flow instability mechanism.

Bye and Babanin conclude their article by stating that there are still many tasks ahead to fully understand wave generation by wind, pointing to the nonlocal aspects of the problem and the link to the contemporary interest lying with the climate system.

Pizzo et al. [8], presents the historical development as well as the present status of the research aiming at explaining the basic mechanics governing the generation of water waves by wind. Miles' work remains a main contribution.

1.1.2 Some Thoughts on the Wave Generation by Wind Problem

A turbulent wind is applied as a premise in many studies. We should note, however, that Miles disregarded the turbulent fluctuations in his stability model, finding them to be of minor importance. This approach seems reasonable and appears generally agreed. It means that the wind has turned unstable and has developed turbulence prior to the instability of the water surface. In this approach, the wind instability is thereby fully decoupled and separated from the water surface instability. Miles also determined the minimum wind required to initiate gravity waves against laminar dissipation of energy in water to be roughly 1 m/s. Thus, Miles' main results are established by energy transfer and energy dissipation considerations. An instability mechanism, such as the GKH instability, works differently.

The GKH model applies vortex sheets and constant flow velocities. This means that the stability of the wind and the stability of the water surface is seen as one and the same problem. Application of a vortex sheet and layers of constant flow for modelling the flow system was criticized by Lord Rayleigh as early as in 1880, see Miles [5, p. 186]. The view that this flow model is an oversimplification of the velocity profile, and therefore invalid, seems to be widely accepted, see, for example, Yih [1, p. 241]. An alternative view might be that the final answer to the validity question on the application of discrete flow layers in stability analysis should rather be given by comparing the theoretical results with the results from good physical measurements.

An inherent problem in the study of wave generation by wind is the fact that from the moment the first waves are generated and are present on the water surface, the process on the water surface is generation and propagation of waves simultaneously. This leads naturally to the approach taken by Miles and others. And it leads further to thinking that there is a coupling between wave generation and wave propagation. This coupling between the two processes is expressed also by Bye and Babanin in their Sect. 1.1.1. However, it seems justified to differentiate between the initial generation of waves on a flat sea surface from the later transfer of wind energy to an existing wave train.

Further, a general agreement seems to exist that a minimum wind velocity is required for the generation of water waves on the water surface. However, the minimum wind question is given some further attention in the following.

In the following study, we establish a simplified flow model for application of the GKH stability model, aiming at seeing if such a simple model may give any results of value to the wind wave problem.

It may be noted that, by the present approach, we formulate the wave generation problem as a stability problem, i.e., the waves are generated by an instability of the water surface, being initially horizontal and flat.

1.2 Objective and Premises for the Study

1.2.1 Study Objective

The objective of the present study is to perform a stability analysis of the wind wave problem by application of the General Kelvin-Helmholtz model as developed in Chap. 1.

1.2.2 Study Premises

An important premise in the present theoretical approach is to represent the boundary layer of the wind at the water surface in modeling the system for stability analysis. The following thinking on the physical system is behind the modeling of the system:

- The wind field does not act directly on the water surface. It has a boundary layer between itself and the water body.
- The water body does not feel the wind field directly but rather the boundary layer of the wind.

Thus, we consider a boundary layer close to the water surface to be a separate part of the flow system. The following specific premises are then established:

- A wind field (WF) is assumed with large depth and with a constant flow velocity U. The air has density ρ_a and is assumed inviscid and incompressible.
- A boundary layer (BL) is assumed to exist between the main wind field and the water body. The boundary layer has a constant velocity εU and a thickness h, where ε is a number smaller than 1. The air density ρ_a in the boundary layer is the same as in the main wind field.
- A water body (WB) is assumed with large depth and is at rest with a horizontal surface. The water has density ρ_w and is assumed inviscid and incompressible.
- The system is in the gravity field.

1.3 Theoretical Flow Model

This leads to defining a model of the flow system as sketched in Fig. 3.1 below. This model will be applied for the following stability analysis.

A vortex sheet (VS 1) separates the wind field and the boundary layer while a second vortex sheet (VS 2) separates the boundary layer and the water body. The wind velocity profile is represented by the uniform velocities U in the wind field and εU in the boundary layer plus an integrated velocity shear included at the two vortex sheets or interfaces. We have thus established a system with two dynamic interfaces for the stability analysis. Development of turbulence in the wind field and generation

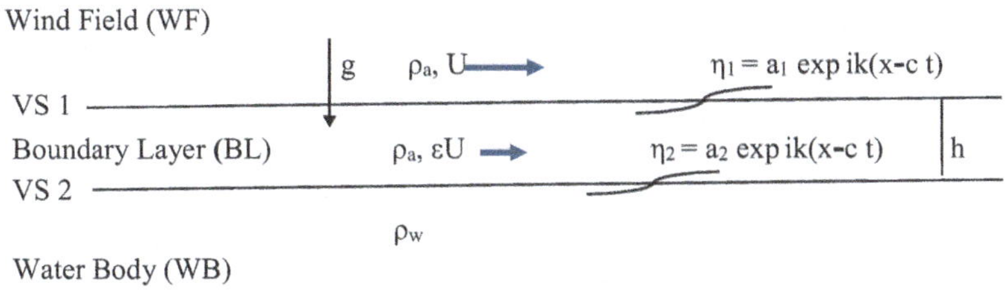

Fig. 3.1 Theoretical flow model of wind field and water body system

of waves on the water surface will thus possibly be the results of the instability character of the whole system.

This system is given a small perturbation as required by the GKH stability model. Different amplitudes on, and a phase shift between, the perturbations on the two interfaces are allowed.

The stability discussions in the following will be performed in terms of the complex phase velocity c. When considering growth rate results, we should keep in mind that the true growth rate is given by the wave frequency $\sigma_i = k\ c_i$.

Modeling the boundary layer as well as the main wind field by flows of constant velocity with an integrated velocity shear at the vortex sheet between them is seemingly a drastic simplification of the total wind field. However, the modelling here should be viewed as a first step, aiming at including the physical effects that are essential for the stability of the system. And it is of course justified by the application of the General Kelvin-Helmholtz stability model.

In the above stability model, turbulence in the wind field is assumed not to exist a priori. This is another simplification in our model. The consequences of our simplifications and the validity of the theoretical model will be discussed in view of the results from the study.

2 Stability Equation

2.1 Application of the Generalized K-H Formula

In the flow system defined above, we introduce a perturbation with wave number k and wave frequency σ as shown. We start by restating the generalized stability equation as given by the result (1.13) in Chap. 1:

$$\rho_{j-1}\left(1/\tanh kh_{j-1} - a_{j-1,j}/\sinh kh_{j-1}\right)\left(c - U_{j-1}\right)^2 + \rho_j\left(1/\tanh kh_j - a_{j+1,j}/\sinh kh_j\right)\left(c - U_j\right)^2 = \left(\rho_j - \rho_{j-1}\right)g/k, \quad j = 1, 2, \ldots, n-1 \tag{3.2}$$

$a_{j-1,j}$ is a short notation for the amplitude ratio a_{j-1}/a_j and similarly for $a_{j+1,j}$. This equation is now applied in turn to the two interfaces, i.e., VS 1 and VS 2 in Fig. 3.1. Thereby the following two homogeneous algebraic equations are established for the amplitudes a_1 and a_2:

$$\left\{\sinh kh(c-U)^2 + \cosh kh(c-\varepsilon U)^2\right\}a_1 - (c-\varepsilon U)^2 a_2 = 0 \tag{3.3}$$

$$\rho_a(c-\varepsilon U)^2 a_1 - \left\{\rho_a \cosh kh(c-\varepsilon U)^2 + \left(\rho_w c^2 - \Delta\rho g/k\right)\sinh kh\right\}a_2 = 0 \tag{3.4}$$

Equation (3.3) is the stability equation for the interface (VS 1) between the main wind field (WF) and the boundary layer (BL) with a last term expressing coupling to the lower interface, i.e., the water surface. Equation (3.4) is likewise the stability equation for the water surface (VS 2) or the interface between the wind boundary layer and the water body (WB). The first term in (3.4) expresses coupling to the wind interface. In Eq. (3.4), we have defined $\Delta\rho = \rho_w - \rho_a$. In the following we apply the definition $\rho_{aw} = \rho_a/\rho_w$. Dividing through Eq. (3.4) by ρ_w the above two equations may now be rewritten as follows:

$$\left\{\sinh kh\,(c-U)^2 + \cosh kh(c-\varepsilon U)^2\right\}a_1 - (c-\varepsilon U)^2 a_2 = 0 \tag{3.5}$$

$$\rho_{aw}(c-\varepsilon U)^2 a_1 - \left\{\rho_{aw}\cosh kh(c-\varepsilon U)^2 + \left[c^2 - (1-\rho_{aw})g/k\right]\sinh kh\right\}a_2 = 0 \tag{3.6}$$

2.2 *Total Eigenvalue Equation*

Equations (3.5) and (3.6) express the stability requirements for the flow system. The determinant of the equations shall be zero to have a nontrivial solution for the amplitudes. This gives the following equation of order four for the phase velocity:

$$\rho_{aw}(c-\varepsilon U)^4 - \left\{\sinh kh(c-U)^2 + \cosh kh(c-\varepsilon U)^2\right\} \cdot \left\{\rho_{aw}\cosh kh(c-\varepsilon U)^2 + \left[c^2 - (1-\rho_{aw})\,g/k\right]\sinh kh\right\} = 0 \tag{3.7}$$

We note that the factor $\rho_{aw} = \rho_a/\rho_w$ is present in three terms in Eq. (3.7). The first term of order four in c is the product of the two coupling terms in Eqs. (3.5) and (3.6). The factor ρ_{aw} is small and comes from the coupling term in Eq. (3.6). The product of the coupling terms must therefore also be a small term. As a first approximation in the following stability analysis, we will disregard the coupling term of Eq. (3.7).

The second term with the factor ρ_{aw} in (3.7) is on a squared velocity term times a wave number function, likely to make this term small. This term is responsible for the instability of the water surface due to the boundary layer velocity. We will consider the effect of it as part of the stability discussions in the following.

The third term in Eq. (3.7) with the ρ_{aw} factor gives a minor correction to the acceleration of gravity. This is without significance and will be disregarded in the following.

2.3 *Simplified Eigenvalue Equation*

Dropping the two ρ_{aw} terms, as said above, we arrive at a simplified stability equation. We divide Eq. (3.7) through by cosh kh to get the new eigenvalue equation expressing the requirement to be satisfied by the phase velocity of the perturbation in the flow system.

$$\left\{\tanh\ kh(c-U)^2 + (c-\varepsilon U)^2\right\} \cdot \left\{\rho_{aw}\ (c-\varepsilon U)^2 + \left(c^2 - g/k\right)\ \tanh\ kh\right\} = 0 \tag{3.8}$$

Solving this equation for the phase velocity c will give the eigenvalue equations, which in turn will be the basis for the stability considerations. We note that Eq. (3.8) expresses the stability of the wind and water system with no coupling effects between the two interfaces. The equation may seem overly simplified. However, we will return to this question in the following sections.

3 WB/BL Interface Instability

3.1 *Eigenvalue Equation*

The second parenthesis of Eq. (3.8) comes from the VS 2 interface, i.e., the water surface, see Eq. (3.6). Requiring this parenthesis to be zero, the eigenvalues of the WB/BL interface are given by the equation:

$$\rho_{aw}(c - \varepsilon U)^2 + (c^2 - g/k)\tanh\ kh = 0 \tag{3.9}$$

We keep the first term for now and solve for the phase velocity c. The following relation results:

$$(\rho_{aw} + \tanh\ kh)c = \rho_{aw}\varepsilon U + i\left\{\rho_{aw}(\varepsilon U)^2 - (\rho_{aw} + \tanh\ kh)\,g/k\right\}^{1/2}(\tanh\ kh)^{1/2} \tag{3.10}$$

3.2 Stability Criterion

From Eq. (3.10), we can establish the main wave parameters. Firstly, we have the propagation rate given by the equation:

$$c_r = \rho_{aw}\varepsilon U/(\rho_{aw} + \tanh\ kh) \tag{3.11}$$

Equation (3.11) shows that the propagation rate is $\rho_{aw}\,\varepsilon U$ for short waves while approaching εU for very long waves.

The growth rate equation is established as follows:

$$c_i = \left\{\rho_{aw}(\varepsilon U)^2 - (\rho_{aw} + \tanh\ kh)g/k\right\}^{1/2} \cdot (\tanh\ kh)^{1/2}/(\rho_{aw} + \tanh\ kh) \tag{3.12}$$

From Eq. (3.12), we may establish the stability criterion, i.e., instability exists for wavelengths satisfying the requirement:

$$\rho_{aw}(\varepsilon U)^2 \geq (\rho_{aw} + \tanh\ kh)gh/kh \tag{3.13}$$

Thus, a high wind velocity U is required to activate this instability, for two reasons:

- The effective velocity of the boundary layer drives the instability and is smaller than the wind velocity by the factor ε.
- The wavelength range stabilized by gravity is moved to very long waves by the small density ratio $\rho_{aw} = \rho_a/\rho_w$.

3.3 Conclusions on the WB/BL Instability

Based on the Sect. 3.2 considerations, we may draw some main conclusions on the instability of the WB/BL interface, i.e., the instability of the water surface due to the flow in the boundary layer:

- The WB/BL instability is strongly influenced by the air-water density ratio strengthening the stabilizing effect of gravity.
- The water surface is unstable due to the flow in the wind boundary layer, as shown by Eq. (3.12). The boundary layer velocity is significantly smaller than the main wind velocity.

Hereby a high wind velocity is required to activate this instability. Thus, the WB/BL instability is not likely to be the mechanism behind normal generation of water waves by wind.

3.4 Final Stability Equation

Based on the above conclusions, we will disregard the first term in the stability equation (3.9), i.e., the remaining ρ_{aw} term in the simplified stability equation (3.8). As a result, the final stability equation for our WF/BL/WB system is the following:

$$\left\{\tanh\ kh\,(c-U)^2+(c-\varepsilon U)^2\right\}\left(c^2-g/k\right)\ \tanh\ kh=0 \tag{3.14}$$

This stability equation expresses that the driving instability in our system is the one at the WF/BL interface, while the perturbations on the water surface will propagate with the deepwater phase velocity given by $c^2 = g/k$. The stability characteristics of this system will be discussed in the following.

4 WF/BL Interface Instability

4.1 WF/BL Interface Eigenvalues

Equation (3.14) has obviously a solution given by setting the first main parenthesis equal to zero. The parenthesis comes from the wind field interface VS 1, see Eq. (3.5). The eigenvalues of the WF/BL interface are then given by the equation:

$$\tanh\ kh\,(c-U)^2+(c-\varepsilon U)^2=0 \tag{3.15}$$

From this, the phase velocity c is follows as:

$$c = U(\tanh\ kh + \varepsilon)/(\tanh\ kh + 1) \\ + i\,\Delta U(\tanh\ kh + \varepsilon)^{1/2}/(\tanh\ kh + 1) \quad (3.16)$$

In this solution for the phase velocity, we disregard the negative imaginary part, since it corresponds to a damped solution, which is not of interest in the stability discussions.

4.2 Main Wave Parameters

From the above Eq. (3.16), we can establish the following main wave parameters:

$$\text{Propagation rate}: \quad c_r = U(\tanh\ kh + \varepsilon)/(\tanh\ kh + 1) \quad (3.17)$$

$$\text{Growth rate}: \quad c_i = \sigma_i/k = \Delta U(\tanh\ kh)^{1/2}/(\tanh\ kh + 1) \quad (3.18)$$

$$\text{Amplitude ratio}: \quad a_2/a_1 = \left\{ \sinh\ kh(c-U)^2 + \cosh\ kh(c-\varepsilon U)^2 \right\}/(c-\varepsilon U)^2 \quad (3.19)$$

The amplitude ratio is established by Eq. (3.5). Hereby we include the last term coupling the perturbation on the wind field interface (VS 1) to the water surface (VS 2). This coupling term in Eq. (3.5) is not small and shall not be disregarded.

4.3 Short Waves

As tanh kh is very close to 1.0 for these waves, assuming $kh \gg 1$, Eqs. (3.17)–(3.19) simplify into the following shortwave equations (3.20) and (3.21). Reference is made to Appendix C.2 for details in developing the amplitude equation (3.22).

$$\text{Propagation rate}: \quad c_r = U(1+\varepsilon)/2 \quad (3.20)$$

$$\text{Growth rate}: \quad c_i = \Delta U/2 \quad (3.21)$$

$$\text{Amplitude ratio}: \quad a_2/a_1 = \exp(-kh). \quad (3.22)$$

This shows that short perturbations on the WF/BL interface are propagated with a velocity that is the average of the velocity of the main wind field and the effective velocity of the boundary layer. The propagation rate is independent of wavelength.

The true growth rate $\sigma_i = k\,c_i$ for short waves is high, i.e., the order of magnitude is kh, which is assumed large ($kh \gg 1$) for a given wind, i.e., given ΔU and h. The true growth rate of the unstable mode is proportional to the wave number, i.e., inversely proportional to the wavelength. The growth rate is also proportional to the velocity difference ΔU between the wind field and the boundary layer. This means that there is no stability limit in the short-wave range.

The above propagation and growth rates both apply to the WF/BL interface and to the WB/BL interface. In the amplitude ratio equation (3.22) kh is by assumption large, i.e., the wavelength is much shorter than the effective thickness of the boundary layer, implying that the negative exponential is very small. The consequence is that the amplitude a_2 on the water surface is much smaller than the amplitude of the perturbation at the WF/BL interface. The function exp $(-kh)$ is the transfer function for the unstable short perturbations generated at the WF/BL interface. The equation shows that unstable perturbations satisfying the shortness requirement $kh \gg 1$, in practical terms, will not reach the water surface. No water waves are generated in this wavelength range and the unstable perturbations will exist in the wind field only.

The water surface response to longer perturbations will be considered in the following.

4.4 *Long Waves*

As tanh kh is very close to kh for these waves, assuming $kh \ll 1$, Eqs. (3.17)–(3.18) simplify into the following long wave equations (3.23)–(3.25) below. Reference is made to Appendix C.3 for the evaluation leading to the amplitude ratio result in Eq. (3.25).

$$\text{Propagation rate:} \quad c_r = \varepsilon U \tag{3.23}$$

$$\text{Growth rate:} \quad c_i = U(1-\varepsilon)(kh)^{1/2} = \Delta U (kh)^{1/2} \tag{3.24}$$

$$\text{Amplitude ratio:} \quad a_2/a_1 = 2(kh)^{1/2}\exp(i\pi/2) + O(kh) \tag{3.25}$$

Equation (3.23) says that long perturbations generated by wind shear against the boundary layer will propagate along both interfaces with the effective velocity of the boundary layer. The fraction ε is not known a priori. By assumption, it is a positive number smaller than 1. We also see that the propagation rate for long waves is independent of the wavelength.

Equation (3.24) shows that the growth rate (in terms of the phase velocity) of long perturbations is low for a given wind, i.e., given ΔU and h, as it goes with $(kh)^{1/2}$. The true growth rate (frequency) goes with $(kh)^{3/2}$, kh assumed small. The long perturbations are thus retarded by their proximity to the water surface, feeling the proximity to it through the boundary layer. The same growth rate applies at both interfaces. As for the short perturbations, the growth rate is proportional to the difference $\Delta U = U(1 - \varepsilon)$ between the velocity of the main wind field and the effective velocity of the boundary layer. By Eq. (3.24), we also see that there is no stability limit in the long wave range, except at kh = 0, which means that there is in fact no stability limit at any finite wavelength.

Equation (3.25) gives the amplitude ratio for long perturbations, as established in Appendix C.3. Thus, we have the relation between the amplitudes on the water surface and at the wind field interface, to a first approximation, given by the equation:

$$a_2 = 2(kh)^{1/2}\exp(i\pi/2)a_1 \quad (3.26)$$

The function $2(kh)^{1/2}\exp(i\pi/2)$ is the transfer function for the long wave instabilities from the WF/BL interface to the water surface. We see that, for long waves, the perturbation amplitude at the water surface is smaller than the amplitude of the perturbation at the WF/BL interface, by the factor $2(kh)^{1/2}$, kh being small.

Another interesting result shown in Eq. (3.26) is that for long waves, the driving wave on the wind field interface is 90° behind the wave on the water surface, i.e., the wind perturbation is in phase with the slope of the generated water wave. Figure 3.2 illustrates the situation.

Miles had this phase difference as a main premise for his theory, see Sect. 1.1.1 and Yih's summary [1]. The air pressure wave, lagging 90° behind the water wave, is clearly an air pressure wave driving the water wave forwards.

It is also interesting, in view of the above result, to consider Miles' statement on his assumption that the important part of the air pressure is in phase with the water wave slope, that this assumption rules out the Kelvin-Helmholtz mechanism of wave generation a priori. Miles' statement is clearly correct in terms of the conventional understanding of the Kelvin-Helmholtz mechanism. However, in terms of the General Kelvin-Helmholtz model, a 90° phase difference is in fact determined by it.

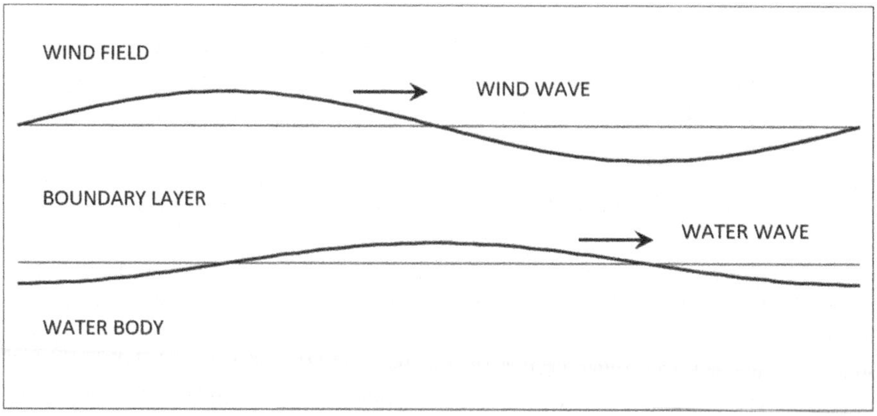

Fig. 3.2 Long wind waves in phase with the water wave slope

5 Wind Effects

5.1 Boundary Layer Assumption

We assume that the boundary layer above the horizontal water surface behaves similarly to the boundary layer above a flat plate. The relation $h \sim \{\nu x/U\}^{1/2}$ is valid for a laminar boundary layer on a flat plate, see, for example, Batchelor [5], section 5.8.

5.2 Effect on Wavelength Intervals

Based on the above assumption, the boundary layer thickness decreases as the wind velocity increases. Thereby, the long wavelength requirement $kh = 2\pi h/\lambda \ll 1$ will include increasingly shorter waves as the wind velocity increases. For example, for a 1 mm thickness of the boundary layer, a 6 cm wavelength gives $kh = 0.1$ approximately, satisfying the longwave requirement. Two consequences of these considerations may be stated as follows:

- For weak winds the results established for short waves will apply.
- For strong winds the boundary layer will be thin and waves satisfying the long wave requirement $kh \ll 1$ will include quite short waves. The long wave results will then be valid for a long range of wavelengths.[1]

5.3 Amplitude Ratio Table

Table 3.1 is established to illustrate a total picture of the absolute value of the amplitude ratio over a wide range of wind velocity and wavelength. Values of the effective thickness h of the boundary layer are given horizontally while wavelength λ, wave number k, and amplitude ratio formulas are given vertically. All lengths are given in meters.

Equation (3.22) for short waves is now formulated as:

$$\mathrm{Abs}(a_2/a_1) = \exp(-kh) \tag{3.27}$$

The amplitude ratio values shown are those near the limit of wave lengths meeting the requirement $kh \gg 1$.

From Eq. (3.26), we may establish the following long wave equation:

[1] The flat plate boundary layer assumption is considered questionable on a water surface where the wave already exists.

Table 3.1 Amplitude ratio a_2/a_1 over a long range of wind velocity and wavelength

Decreasing Wind Velocity →						
λ↓	h→	0.001	0.003	0.005	0.01	0.03
0.001	kh ↘	6.28		***Short*** *perturbations Finite growth rate Vanishing water wave amplitude*		
	$e^{(-kh)}$ ↘	**1.87 E-3**				
0.003	"		6.28			
	"		**1.87 E-3**			
0.005	"	~ 1.0		6.28		
	"	**~ 0.37/2**		**1.87 E-3**		
0.01	"		↖		6.28	
	"		***Intermediate*** *perturbations First water waves*		**1.87 E-3**	
0.03	kh ↘	0.209				6.28
	$2(kh)^{1/2}$ ↘	**0.91**		↘		**1.87 E-3**
0.05	"	0.1256	0.377		~ 1.0	
	"	**0.71**	**1.23**		**~ 0.37/2**	
0.10	"	0.0628	0.1884	0.314		~ 1.0
	"	**0.50**	**0.87**	**1.12**		**~ 0.37/2**
0.50	"	***Long*** *perturbations Retarded growth Water waves are generated*			0.1256	
	"				**0.71**	
1.0	"				0.0628	0.1884
	"				**0.50**	**0.87**
5.0	"	0.001256	0.00377	0.00628	0.01256	0.0377
	"	**0.071**	**0.12**	**0.16**	**0.22**	**0.39**
10.0	"	0.000628	0.001884	0.00314	0.00628	0.01884
	"	**0.050**	**0.087**	**0.11**	**0.16**	**0.27**

$$\mathrm{Abs}(a_2/a_1) = 2(kh)^{1/2} \tag{3.28}$$

Actual absolute values of the amplitude ratio for short and long waves are shown. The upper right part of Table 3.1 is not shown as the actual short-wave numbers are vanishingly small.

The case kh = 1.0 in the intermediate range of wavelength is included as a suggested value for the first waves, see Sect. 5.4.2 below. The value kh = 1.0 corresponds nearly, but not exactly, to the shown values of λ and h. The amplitude ratio, according to the short wave and long wave equation, is shown in the line below it. As the shortwave value 0.37 is quite low while the long wave value of 2 may be a bit high, the true amplitude ratio for the first wave may be somewhere between these two values. However, if the long wave equation applies for kh = 1, the amplitude ratio of 2 may apply. Two interesting observations are made in Table 3.1:

- The amplitude ratio has a maximum in the intermediate range of waves.
- The first wave is longer for a weaker wind.

5.4 *Instability Characteristics*

5.4.1 Progressive Instability

Based on the above stability results as well as Table 3.1, the following considerations may be made:

- Short perturbations, $kh \gg 1$, are most unstable, i.e., having the highest growth rate. The shortest perturbations are not seen on the water surface. These short perturbations will develop first and will, expectedly, develop wind turbulence.
- Long perturbations, $kh \ll 1$, with a reduced growth rate, generate waves on the water surface by the transfer function $2(kh)^{1/2}\exp(i\pi/2)$.
- Since water waves are not generated by the short perturbations, with a high growth rate, while water waves are generated by the long wind perturbations, having a low growth rate. This indicates that the instability develops progressively, i.e., developing gradually from short towards longer waves.

5.4.2 First Wave

Assuming a progressive character of the instability, the following considerations may be added:

- An intermediate range of unstable waves exist between the short and the long waves. The first waves to be seen on the water surface should be the first wave to experience reduced growth rate due to its proximity to the water surface.
- The following criterion is proposed to determine the first wave being generated by the air flow:

$$kh = 1 \tag{3.29}$$

The first wave, as determined by this criterion, will have a length given by: $\lambda = 2\pi h$. This criterion for determining the first wave is not, as we can see, an analytical deduction from the stability analysis but rather a criterion established by a simple judgement. The judgement goes as follows: Waves developing progressively will, at a wavelength of 6.28 times the thickness of the boundary layer thickness, experience proximity to the water surface. Thereby the growth rate is reduced, and the instability will be transferred to the water surface as a more slowly growing wave.

The progressive nature of the instability is a question to be addressed by future work including laboratory testing. Some main objectives should be in focus:

- Verification of the progressive nature of the instability.
- If the progressive character is confirmed for short and intermediate waves: Verification that the progressive development continues into and through the long wave regime.
- Verification of the first wave criterion.

If the first wave criterion is confirmed, the criterion may then be turned around: Measuring the first wave by length, the effective thickness h of the boundary layer is determined by the relation: $h = \lambda/2\pi$.

5.5 A First Wave Demonstration

5.5.1 Two First Wave Characteristics

Two characteristics, based on the above observations in Table 3.1, are now stated for the first water waves to be generated by an air flow:

- We assume that the GKH instability generating water waves develops progressively. According to Table 3.1, the water wave amplitude in ratio to the wave at the wind interface seems to maximize for the first wave to be generated. This indicates that the first wave may be dominating on the water surface for a given air flow.
- Table 3.1 expresses that the wavelength of the first wave will be longer if the wind velocity is reduced.

A simple demonstration was set up to check the above effect on the wavelength of the first waves generated by an air flow of different velocities.

5.5.2 Demonstration Setup

A small flume, about 50 cm long, 25 cm wide, and 8 cm deep, was filled with water up to a depth of about 3 cm. The water surface was at rest initially.

The demonstration applied a thin plastic pipe, ID = 6 mm, to blow air on to the water surface. This simplifies the wind field to an air flow meeting the water surface at one location at the end of the flume. The following thinking is behind this setup:

- The same wave generating mechanism acting between the wind field and the water surface will be acting locally also in the test flume where the air flow from the pipe meets the water surface.
- By use of the pipe, waves will be generated on the water surface just below the pipe end where no propagating waves are present. The generated waves will propagate away from this location. Hereby the wave generation is separated from the wave propagation.

- The boundary layer thickness increases when the wind velocity decreases.
- The criterion for the first wave, as formulated by Eq. (3.29), may be considered as tentatively established.

The above indications should of course be verified by future work.

5.5.4 Minimum Wind

According to the theoretical results above, there is no minimum wind generating instability at the WF/BL interface. However, the wave generating mechanism, as described, is such that the boundary layer thickness increases as the wind velocity decreases. The consequence is that the length of the generated first water waves—the waves of highest amplitude ratio—increases for a decreasing air flow velocity. At the same time, the water wave amplitude will be reduced as less energy is available in the wind. By this mechanism—increasing wavelength and decreasing amplitude—the wave generation will gradually die out when the wind velocity is reduced. An in-detail description of this process is not given here. Thus, the question of the minimum wind remains and should be addressed in future work.

6 Result Summary

A limited study, as the present one, does of course not allow any far-reaching conclusions. However, the most interesting points from the work may be summarized as follows:

- The WF/BL/WB system has two General KH Instability interfaces:
  - The water surface with a boundary layer above (WB/BL)
  - The wind field shear layer with a boundary layer below (WF/BL)
- The WB/BL instability acting at the water surface is subject to two weakening effects:
  - A strong stabilizing effect by gravity, strengthened by the water/air density ratio
  - A velocity reduction from the main wind field to the boundary layer
- The WF/BL instability acting at the wind field interface is the driving instability generating waves on the water surface. This instability is subject to the following characteristics:
  - There is no stabilizing effect by gravity. All wavelengths are unstable.
  - Short wind perturbations, $kh \gg 1$.

    Have a finite growth rate, i.e., they are rather strongly unstable, expectedly generating wind turbulence.

The short water wave amplitude is strongly reduced by the factor exp (−kh), meaning that no water waves will be generated in the short-wave range.

- Long wind perturbations, $kh \ll 1$

  Experience proximity to the water surface, giving a retarded instability.
  Long waves in the wind field are transferred through the boundary layer to the water surface where they generate water waves.
  Long wind waves are in phase with the slope of the water wave, i.e., the water wave is 90° ahead of the wind wave.
  The long water wave amplitude is reduced by a factor $2(kh)^{1/2}$ on the amplitude of the wind field perturbation.

- The WF/BL instability is likely to be progressive. This should be confirmed by future work.
- The length of the first waves seen on the water surface generated by the air flow is tentatively determined by the condition $kh = 1$.
- The wavelength of the first waves increases, while the amplitude of the first waves will be reduced when the air flow velocity decreases.

7 Some Final Remarks

Two remarks are made on the preceding stability analysis:

- In Sect. 1.3, we pointed to the assumption in the theoretical model that wind turbulence is not existing a priori. In our physical demonstration, the wind is replaced by a turbulent jet from the pipe. There is no indication in the present work that the turbulence should play an important role. This is in line with Miles' assumption that turbulence is of minor importance.
- An obvious weakness of the present approach is the missing knowledge of the boundary layer in this system. The assumption made on it, that it behaves like the flat plate boundary layer, seems to be confirmed by our simple demonstration. However, this behavior will be highly questionable on a water surface with propagating waves. Despite this weakness, the present work, based on the GKH stability model, seems to offer some interesting results and possibly a new approach to this important problem.

There are certainly many questions left unanswered by the present stability results. The present work may be seen as a short initial step. Thus, verification work as well as improvements and extensions of the present work are strongly recommended. Several topics to be addressed in future work are pointed out in the preceding text. High quality physical testing is the way to qualify this type of theoretical results and will be an essential part of the work to be performed on this important problem in the future.

References

1. Yih, Chia-Shun. 1980. *Stratified Flows*. New York: Academic Press.
2. Lamb, Horace. 1932. *Hydrodynamics*. Cambridge: Cambridge University Press.
3. Bye, J.A.T., and A.V. Babanin. 2009. Wave generation by wind. In *Encyclopedia of Ocean Sciences*, 2nd ed. San Diego: Academic Press.
4. Batchelor, G.K. 1970. *An Introduction to Fluid Dynamics*. Cambridge: Cambridge University Press.
5. Miles, J.W. 1957. On the generation of surface waves by shear flows. *Journal of Fluid Mechanics* 3: 185.
6. Miles, J. W. 1959. On the generation of surface waves by shear flows. Part 3. Kelvin-Helmholtz instability. *Journal of Fluid Mechanics* 6: 583.
7. Ursell, F. 1956. Wave generation by wind. In *Surveys in Mechanics*, 216. Cambridge: Cambridge University Press.
8. Pizzo, N., L. Deike, and A. Ayet. 2021. How does the wind generate waves? *Physics Today* 74 (11): 38.

$$\varepsilon = (kh_2)^{1/2} \tag{A.4}$$

The eigenvalue expansion is formulated by the series:

$$\xi = \xi_0 + \xi_1\varepsilon + \xi_2\varepsilon^2 + \xi_3\varepsilon^3 + \cdots \tag{A.5}$$

s the finite eigenvalue in this development. The higher powers of ξ are derived as ows:

$$\begin{aligned} \xi^2 &= \xi_0{}^2 + 2\xi_0\,\xi_1\varepsilon + (2\,\xi_0\,\xi_2 + \xi_1{}^2)\varepsilon^2 + O(\varepsilon^3) \\ \xi^3 &= \xi_0{}^3 + 3\xi_0{}^2\xi_1\varepsilon + 3\xi_0(\xi_0\,\xi_2 + \xi_1{}^2)\varepsilon^2 + O(\varepsilon^3) \\ \xi^4 &= \xi_0{}^4 + 4\xi_0{}^3\xi_1\varepsilon + 2\,\xi_0{}^2(2\,\xi_0\,\xi_2 + 3\xi_1{}^2)\varepsilon^2 + O(\varepsilon^3) \end{aligned} \tag{A.6}$$

By this, the eigenvalue equation (A.2) is developed into the following asymptotic ation:

$$\begin{aligned} &\xi_0{}^2\{\alpha\,\xi_0{}^2 - 2\,\alpha\xi_0 + \alpha + \beta\} + \xi_0\,\xi_1\{4\alpha\,\xi_0{}^2 - 6\,\alpha\,\xi_0 + 2(\alpha + \beta)\}\varepsilon \\ &+\{2\,\alpha\,\xi_0{}^2(2\,\xi_0\xi_2 + 3\,\xi_1{}^2) - 6\,\alpha\,\xi_0(\xi_0\xi_2 + \xi_1{}^2) + (\alpha + \beta)(2\,\xi_0\xi_2 + \xi_1{}^2)\}\varepsilon^2 \\ &= \{\gamma\,\xi_0{}^4 - 4\gamma\,\xi_0{}^3 + (6\gamma + \delta)\,\xi_0{}^2 - 2(2\gamma + \delta)\xi_0 + \gamma + \delta + \kappa\}\varepsilon^2 + O(\varepsilon^3) \end{aligned} \tag{A.7}$$

The first term in the above expansion is of order ε^0 and determines the possible ite solutions for ξ:

$$\xi_0{}^2\{\alpha\xi_0{}^2 - 2\alpha\,\xi_0 + \alpha + \beta\} = 0 \tag{A.8}$$

Here a double root is:

$$\xi_0(3,4) = 0 \tag{A.9}$$

t giving any finite eigenvalues. The two remaining roots of (A.8) are given by:

$$\alpha\,\xi_0{}^2 - 2\alpha\,\xi_0 + \alpha + \beta = 0 \tag{A.10}$$

$$\xi_0{}^{(1,2)} = 1 \pm (-\beta/\alpha)^{1/2} \tag{A.11}$$

Proceeding further in the development of the finite eigenvalues, we would find at there is no correction term of order $(kh_2)^{1/2}$, while a correction term of order kh_2 ll exist. The details are given in the Appendix A of Ref. [3] of Chap. 3.

Epilogue

The General Kelvin-Helmholtz (GKH) stability model, as established in Chap. 1, is the reason why this book is written. All the other material in this book rests on it. Some of the results that have been established by its applications seem quite remarkable and give an opportunity for some further thinking about the GKH stability model.

The GKH model applies vortex sheets in combination with uniform flow fields of constant velocity for modeling of the flow system. The vortex sheet concentrates the flow shear at the interface, thereby establishes a discontinuity in the tangential velocity. Additionally, the density field is discretized by the same sheet or interface. As we have pointed out earlier, a wide agreement seems to exist, saying that this discretization of the flow field is overly simplified and therefore invalid, see, for example, the book by Yih, p. 241, see Ref. [1] of Chap. 3. Most studies, on the sand wave problem as well as on the wind wave problem, over many decades, apply velocity profiles with distributed velocity shear. Our studies have demonstrated that the GKH stability model, being based on "overly simplified" velocity and density fields, leads to results that are surprisingly in line by the physical behavior of the actual flow systems. This may particularly be said for the sand wave study and demonstrates the usefulness of the approach based on discrete flow layers separated by vortex sheets, at least as a first step in studying these phenomena. Thus, in view of the present results, there may be a reason to reconsider the usefulness and validity of applying discretized velocity and density fields in such stability studies.

Some common features of Chaps. 2 and 3 are striking. Firstly, it should be pointed out that the GKH instability seems to develop progressively in both cases. The progressive character is conclusive for the sand wave case, and there are good indications also for the wind wave case. Whether the progressive character is valid for the GKH instability in general is an important question to be addressed in future work.

An important feature of the GKH instability is that it allows dynamic interaction between dynamic faces, i. e. interfaces and also the free surface. The sand wave

H. Moshagen, *The General Kelvin-Helmholtz Stability Model*,
https://doi.org/10.1007/978-3-031-61984-7

study shows this clearly by the influence of the free surface on the stability of the interface and the bed in the long wave range.

Another important feature of the GKH mechanism is its ability to transfer the instability through a bed layer and through a boundary layer. The sand wave study demonstrates that the instability is transferred from the bed layer interface to the sand bed. The wind wave study demonstrates that the unstable long perturbations in the wind field act through the boundary layer and generates water waves on the water surface, i. e. a transfer without coupling effects. Both studies, in Chaps. 2 and 3, are examples where the bed layer in channel flow and the boundary layer in the wind field play key roles. The bed and boundary layers determine the small eigenvalues of the system which in turn determine the stability behavior of the systems for long waves. Both sand waves on the bed and wind waves on the water surface are, according to the present studies, long wave instabilities driven by the main flows above.

Another common feature is this: our studies give reason to understand the type of instability generating waves as different from instabilities generating turbulence by the difference of their growth rate. The wave generations we have studied are in both cases the result of a long wave retarded GKH instability, i.e., an instability of long waves being given a reduced growth rate by their proximity to the sand bed or to the water surface below. Turbulence from the KH instability, on the other hand, results, expectedly, from the high growth rate of short perturbations in both cases.

As already remarked on the results of the sand wave study in Chap. 2, the conclusions from the present study indicate that the earlier limited success, by applying the conventional KH model to describe these well-known instabilities, may be due to the undeveloped status of the conventional KH model. The conventional KH model implies use of a one "degree of freedom" model, i.e., two layers with one interface, for the stability studies. When more layers make a meaningful flow model for an actual stability problem, the GKH model applies and does not have this limitation.

Our performed studies in Chaps. 2 and 3 show us that the GKH model takes us quite a long step forward from the conventional KH model. In future work, a natural first step in stability studies of stratified flow systems should be to apply the GKH model, followed eventually by more detailed studies based on the results from the GKH model. The GKH stability model is, by the present work, considered well established for this type of use.

Appendix A: Generation of Sand Wa in Alluvial Channels—Asymptotic E of Eigenvalues

A.1 Finite Eigenvalues

We start by stating the eigenvalue equation (2.12):

$$\left\{\tanh\ kh_1/kh_1 - F^2(\xi-1)^2\right\}(r+1)F^2\tanh\ kh_1\xi^2$$
$$=\left\{F^4\tanh^2 kh_1(\xi-1)^4-(r+1)F^2(\tanh\ kh_1/kh_1)(\xi\right.$$
$$\left.+r\ (\tanh\ kh_1/kh_1)^2\right\}kh_2+O\left\{(kh_2)^2\right\}$$

We rewrite this equation in a more tractable form by:

$$\left\{\alpha(\xi-1)^2+\beta\right\}\xi^2=\left\{\gamma(\xi-1)^4+\delta(\xi-1)^2+\kappa\right\}kh_2+O\{$$

Here the new symbols are defined as follows:

$$\alpha=(r+1)F^4\tanh\ kh_1$$
$$\beta=-(r+1)F^2\tanh^2 kh_1/kh_1$$
$$\gamma=-F^4\tanh^2 kh_1$$
$$\delta=(r+1)F^2\tanh\ kh_1/kh_1$$
$$\kappa=-r(\tanh\ kh_1/kh_1)^2$$

Referring to Sect. 6.1 in Chap. 2, we define the expansion para

H. Moshagen, *The General Kelvin-Helmholtz Stability Model*,
https://doi.org/10.1007/978-3-031-61984-7

The result for the finite eigenvalues is now given by entering in (A.11) the expressions of α and β, defined in (A.3), and reads as follows:

$$\xi_0(1,2) = F^{-1}\left\{ F \pm (\tanh\, kh_1/kh_1)^{1/2} \right\} + O\left\{ (kh_2)^{1/2} \right\} \tag{A.12}$$

A.2 Small Eigenvalues

The determination of the small eigenvalues of Eq. (A.1) is most effectively done by disregarding the finite eigenvalues. Hereby we perform the asymptotic expansion by the series:

$$\begin{aligned} \xi &= \xi_1\varepsilon + \xi_2\varepsilon^2 + \xi_3\varepsilon^3 + \xi_4\varepsilon^4 + \cdots \\ \xi^2 &= \xi_1{}^2\varepsilon^2 + 2\,\xi_1\xi_2\varepsilon^3 + \left(\xi_2{}^2 + 2\,\xi_1\xi_3\right)\varepsilon^4 + \cdots \\ \xi^3 &= \xi_1{}^3\varepsilon^3 + 3\,\xi_1{}^2\xi_2\varepsilon^4 + \cdots \\ \xi^4 &= \xi_1{}^4\varepsilon^4 + \cdots \end{aligned} \tag{A.13}$$

As before we define the expansion parameter by: $\varepsilon = (kh_2)^{1/2}$. Development of the rewritten eigenvalue equation (A.2) now gives:

$$\begin{aligned} &(\alpha+\beta)\,\xi_1{}^2\varepsilon^2 + \left\{ -2\,\alpha\,\xi_1{}^2 + 2(\alpha+\beta)\xi_2 \right\}\xi_1\varepsilon^3 = (\gamma+\delta+\chi)\,\varepsilon^2 \\ &\quad - 2(2\gamma+\delta)\xi_1\varepsilon^3 + O\{\varepsilon^4\} \end{aligned} \tag{A.14}$$

The ε^2 terms now determine a first approximation to the small eigenvalues by

$$(\alpha+\beta)\xi_1{}^2 = (\gamma+\delta+\chi) \tag{A.15}$$

or

$$\xi_1{}^2 = (\gamma+\delta+\chi)/(\alpha+\beta) \tag{A.16}$$

When the definitions in (A.3) are introduced, we have the following result:

$$\begin{aligned}\xi_1^{(3,4)} &= \pm i\left\{F^4 - (r+1)F^2\coth kh_1/kh_1 + r/(kh_1)^2\right\}^{1/2} \\ &\cdot\{(r+1)(F^2 - \tanh kh_1/kh_1)F^2\coth kh_1\}^{-1/2} \\ &= \pm i\{F^2 - f_1(kh_1,r)\}^{1/2}\{F^2 - f_2(kh_1,r)\}^{1/2} \\ &\cdot\{(r+1)(F^2 - f_3(kh_1))\}^{-1/2}\{F^2\coth kh_1\}^{-1/2}\end{aligned} \tag{A.17}$$

Here we have introduced the functions:

$$\begin{aligned} f_1(kh_1,r) &= ½(r+1)\coth kh_1/kh_1 \\ &+\left\{(r+1)(\coth kh_1/kh_1)^2/4 - r/(kh_1)^2\right\}^{1/2}\end{aligned} \tag{A.18}$$

$$\begin{aligned} f_2(kh_1,r) &= ½(r+1)\coth kh_1/kh_1 \\ &-\left\{(r+1)^2(\coth kh_1/kh_1)^2/4 - r/(kh_1)^2\right\}^{1/2}\end{aligned} \tag{A.19}$$

$$f_3(kh_1) = \tan kh_1/kh_1 \tag{A.20}$$

This eigenvalue solution may be purely imaginary or real depending on the variables in the brackets.

A preliminary result for the small eigenvalues now reads as follows:

$$\begin{aligned}\xi^{(3,4)} &= \pm i\xi_i^{(3,4)} = \pm i\,\{F^2 - f_1(kh_1,r)\}^{1/2}\{F^2 - f_2(kh_1,r)\}^{1/2} \\ &\cdot\{(r+1)(F^2 - f_3(kh_1))\,F^2\coth kh_1\}^{-1/2}(kh_2)^{1/2} + O\{kh_2\}\end{aligned} \tag{A.21}$$

This solution is not valid at and close to the relation:

$$F^2 = f_3(kh_1) = \tanh\ kh_1/\,kh_1, \tag{A.22}$$

since it is a singularity condition in (A.21). The eigenvalue solution at $F^2 = \tanh kh_1/kh_1$ is considered separately in Sect. A.3.

Proceeding to the term of order ε^3 in (A.14), we get the solution:

$$\xi_2 = \{\alpha(\chi - \gamma) - \beta(2\gamma + \delta)\}\,(\alpha + \beta)^{-2} \tag{A.23}$$

By use of the definitions (A.3), we arrive at the solution

$$\begin{aligned}\xi_2^{(3,4)} &= \left\{F^4 - 2\,F^2 \tanh\, kh_1/kh_1 + (kh_1)^{-2}\right\} \\ &\cdot\left\{(F^2 - \tanh\, kh_1/kh_1)^2 (r+1)\coth\, kh_1\right\}^{-1} \\ &= \left\{(F^2 - \tanh\, kh_1/kh_1)^2 + (1 - \tanh^2 kh_1)(kh_1)^{-2}\right\} \\ &\cdot\left\{(F^2 - \tanh\, kh_1/kh_1)^2 (r+1)\coth kh_1\right\}^{-1}\end{aligned} \tag{A.24}$$

It is easily seen that both eigenvalues are real and positive expressing the same downstream propagation rate of the unstable mode. The complete solution to an order of magnitude up to kh_2 is now established for small eigenvalues and may be written as follows:

$$\begin{aligned}\xi^{(3,4)} &= \pm i\left\{F^2 - f_1(kh_1, r)\right\}^{1/2}\left\{F^2 - f_2(kh_1, r)\right\}^{1/2} \\ &\cdot\left\{(r+1)(F^2 - f_3(kh_1))\,F^2 \coth\, kh_1\right\}^{-1/2}(kh_2)^{1/2} \\ &+\left\{(F^2 - \tanh\, kh_1/kh_1)^2 + (1 - \tanh^2 kh_1)(kh_1)^{-2}\right\} \\ &\left\{(F^2 - \tanh\, kh_1/kh_1)^2 (r+1)\coth kh_1\right\}^{-1}(kh_2) + O\left\{(kh_2)^{3/2}\right\}\end{aligned} \tag{A.25}$$

A.3 Eigenvalues at $F^2 = \tanh kh_1/kh_1$

This case is investigated by starting from the following rewritten form of the eigenvalue equation (A.1) where the condition $F^2 = \tanh kh_1/kh_1$ has been applied:

$$\alpha(\xi - 2)\xi^3 = \left\{\gamma(\xi - 1)^4 + \delta(\xi - 1)^2 + \chi\right\}kh_2 \tag{A.26}$$

The parameters in this equation are defined as follows:

$$\begin{aligned}\alpha &= (r+1)\ \tanh\ kh_1 \\ \gamma &= -\tanh^2 kh_1 \\ \delta &= (r+1) \\ \chi &= -r\end{aligned} \tag{A.27}$$

Inspection of (A.26) gives the finite eigenvalue solution:

$$\xi^{(1)} = 2 + O(kh_2) \tag{A.28}$$

This is seen to be the same solution as the one (positive root) found in (A.12).

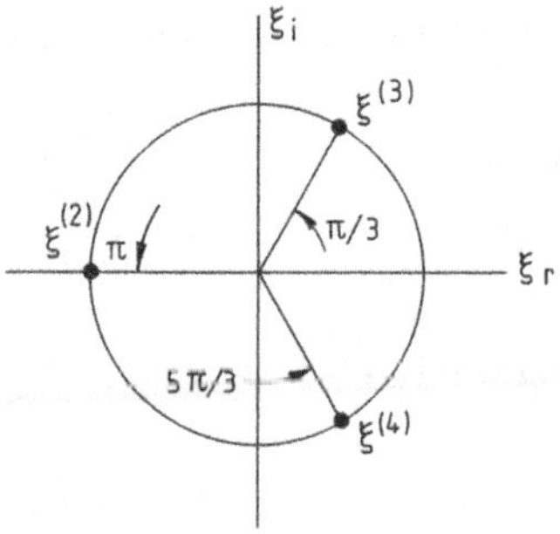

Fig. A.1 Position of the three small eigenvalues in the Argand diagram

Inspection of (A.26) also leads to the following perturbation parameter for the small eigenvalues:

$$\varepsilon = (kh_2)^{1/3} \tag{A.29}$$

We are still requiring $kh_2 \ll 1$ in search for these eigenvalues. They are now searched by developing ξ follows:

$$\xi = \xi_1\varepsilon + \xi_2\varepsilon^2 + \xi_3\varepsilon^3 + \cdots \tag{A.30}$$

The powers ξ^2, ξ^3, and ξ^4 are established as in (A.1).
After introduction of these in (A.1), we get the following eigenvalue equation:

$$\begin{aligned} &-2\alpha\,\xi_1{}^3\varepsilon^3 + \alpha\,\xi_1{}^2(\xi_1{}^2 - 6\xi_2)\varepsilon^4 \\ &= (\gamma + \delta + \chi)\,\varepsilon^3 - 2(2\gamma + \delta)\,\xi_1\,\varepsilon^4 + O\{\varepsilon^5\} \end{aligned} \tag{A.31}$$

The terms of order ε^3 now give the solution:

$$\xi_1{}^3 = -(\gamma + \delta + \chi)(2\alpha)^{-1} \tag{A.32}$$

Introducing definitions (A.27) and some manipulation of terms we arrive at the eigenvalue solution:

$$\begin{aligned} \xi^{(n+2)} &= \exp\{i(3 - 2n)\pi/3\}\{(r+1)\sinh 2kh_1\}^{-1/3}(kh_2)^{1/3} \\ &\quad + O\left\{(kh_2)^{2/3}\right\}, \quad n = 0, 1, 2 \end{aligned} \tag{A.33}$$

The eigenvalues to first order of approximation clearly have the same absolute value:

$$\mathrm{Abs}\left\{\xi^{(n+2)}\right\} = \{(r+1)\sinh 2kh_1\}^{-1/3}(kh_2)^{1/3}, \quad n=0,1,2 \tag{A.34}$$

Their arguments have the following values and position in the Argand diagram (Fig. A.1):

$$\mathrm{Arg}\left\{\xi^{(n+2)}\right\} = \{\pi/3, \pi, 5\pi/3\}. \tag{A.35}$$

The three small eigenvalues are now written as follows, separating their real and imaginary parts:

$$\begin{gathered}\xi_r^{(2)} = -\{(r+1)\sinh 2kh_1\}^{-1/3}(kh_2)^{1/3} \\ \xi_i^{(2)} = 0\end{gathered} \tag{A.36}$$

$$\xi_r^{(3)} = ½\{(r+1)\sinh 2kh_1\}^{-1/3}(kh_2)^{1/3}\xi_i^{(3)} = ½(3)^{1/2}\{(r+1)\sinh 2kh_1\}^{-1/3}(kh_2)^{1/3} \tag{A.37}$$

$$\begin{gathered}\xi_r^{(4)} = \xi_r^{(3)} \\ \xi_i^{(4)} = -\xi_i^{(3)}\end{gathered} \tag{A.38}$$

The eigenvalue solution $\xi^{(3)}$ has a positive imaginary part. Thus, it is the one governing the instability. The final eigenvalue $\xi^{(4)}$ is the damped solution with the same real and (negative) imaginary parts as the unstable eigenvalue $\xi^{(3)}$.

Appendix B: Generation of Sand Waves in Alluvial Channels—Development of Amplitude Ratios

B.1 Amplitude Ratio for Finite Eigenvalues

Sections, equations, and formulas or references pointing back in the book within the Appendix B text all point to Chap. 2. We start by restating the finite eigenvalue equation (2.12) as follows:

$$F^2(\xi - 1)^2 - \tanh\, kh_1/kh_1 = O(kh_2) \tag{B.1}$$

We also state the equation for the dynamic condition at the free surface, by Eq. (2.1),

$$(1/\tanh\, kh_1 - a_{21}/\sinh\, kh_1)(c - U_1)^2 = g/k \tag{B.2}$$

Here a_{21} is a short notation for the amplitude ratio a_2/a_1, where a_1 is the amplitude at the free surface and a_2 is the amplitude at the bed interface, ref. Fig. 2.2. This equation is reformulated by the definitions in Sects. 5.2 and 5.3 in Chap. 2, restated here:

$$\begin{gathered} \zeta = c - U_2 \\ r = \Delta\rho/\rho_1 = (\rho_2 - \rho_1)/\rho_1 \\ \xi = \zeta/\Delta U = \zeta/(U_1 - U_2) \\ F^2 = \Delta U^2/g\, h_1 \end{gathered}$$

Equation (B.2) then takes the form:

H. Moshagen, *The General Kelvin-Helmholtz Stability Model*,
https://doi.org/10.1007/978-3-031-61984-7

$$a_{21} = \left\{F^2(\xi - 1)^2 - \tanh\ kh_1/kh_1\right\}F^{-2}\cosh\ kh_1(\xi - 1)^{-2} \tag{B.3}$$

From the two Eqs. (B.1) and (B.3), we get the result

$$a_{21} = O(kh_2) \tag{B.4}$$

According to this result, the amplitude of the perturbation at the bed layer interface is an order of magnitude smaller than the amplitude at the free surface for the finite eigenvalues.

The equation for the amplitude ratio given by the dynamic condition at the bed layer interface may be used as well, giving the same result.

B.2 Amplitude Ratio for Small Eigenvalues

The case of small eigenvalues is treated in a similar way as the case of finite eigenvalues above. We start by restating the amplitude ratio equation (B.3), which is based on the dynamic condition at the free surface:

$$a_{21} = \left\{F^2(\xi - 1)^2 - \tanh\ kh_1/kh_1\right\}F^{-2}\cosh\ kh_1(\xi - 1)^{-2} \tag{B.5}$$

Developing the two parentheses $(\xi-1)^2$ and $(\xi-1)^{-2}$ for small values of ξ, i.e., $\xi = O\{(kh_2)^{1/2}\}$, we get the following result for the amplitude ratio:

$$a_{21} = \{F^2 - \tanh\ kh_1/kh_1\}F^{-2}\cosh\ kh_1 + O\left\{(kh_2)^{1/2}\right\} \tag{B.6}$$

The premise for the eigenvalue solution applied here is that it is not valid close to or at the transition condition $F^2 - \tanh\ kh_1/kh_1$. The same restriction applies consequently to the amplitude ratio.

B.3 Amplitude Ratio at $F^2 = \tanh\ kh_1/kh_1$

The one finite eigenvalue is included in the result (B.4). For the further three eigenvalue cases, we therefore apply the result given in Eq. (A.8). The amplitude ratio equation is based on the dynamic condition at the free surface, as given by (B.3). This is now rewritten as follows:

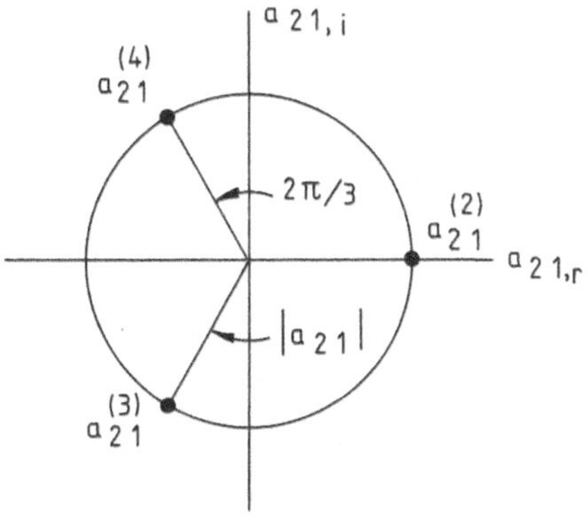

Fig. B.1 Position of the three amplitude ratios in the Argand diagram

$$a_{21} = \left\{F^2(\xi - 1)^2 - \tanh\ kh_1/kh_1\right\}F^{-2}\cosh\ kh_1(\xi - 1)^{-2} \\ = -2\cosh\ kh_1\,\xi + O(\xi^2) \tag{B.7}$$

Since $\xi = O\{(kh_2)^{1/3}\}$, ξ^2 will be $O\{(kh_2)^{2/3}\}$. Introducing now the eigenvalues from (A.33), we get the three corresponding amplitude ratios:

$$a_{21}{}^{(n+2)} = 2\cosh\ kh_1\{(r+1)\sinh 2kh_1\}^{-1/3}\exp\{-i2n\,\pi/3\}(kh_2)^{1/3} \\ +O\left\{(kh_2)^{2/3}\right\},\quad n = 0, 1, 2 \tag{B.8}$$

As we see, all three amplitude ratios are of order $(kh_2)^{1/3}$.

The three amplitude ratios, to first order of approximation, have the same absolute value:

$$\mathrm{Abs}\left\{a_{21}{}^{(n+2)}\right\}_{n=0,1,2} = 2\cosh\ kh_1\{(r+1)\ \sinh 2kh_1\}^{-1/3}(kh_2)^{1/3} \tag{B.9}$$

Their arguments are, by values and position in the Argand diagram, the following (Fig. B.1):

$$\mathrm{Arg}\left\{a_{21}{}^{(n+2)}\right\}_{n=0,1,2} = \{0,\ -2\pi/3, 2\pi/3\} \tag{B.10}$$

Appendix C: Wave Generation by Wind—Development of Amplitude Ratios

C.1 Method

We restate for further consideration Eq. (3.5) determining the amplitude ratio for the two interfaces considered in the foregoing WF/BL stability analysis:

$$\left\{ \sinh\ kh(c - U)^2 + \cosh\ kh(c - \varepsilon U)^2 \right\} a_1 - (c - \varepsilon U)^2 a_2 = 0 \qquad \text{(C.1)}$$

Solving for the amplitude ratio, we get

$$a_2/a_1 = \left\{ \sinh\ kh(c - U)^2 + \cosh\ kh(c - \varepsilon U)^2 \right\} / (c - \varepsilon U)^2 \qquad \text{(C.2)}$$

or

$$a_2/a_1 = \left\{ (\underline{c} - 1)^2 / (\underline{c} - \varepsilon)^2 \right\} \sinh\ kh + \cosh\ kh \qquad \text{(C.3)}$$

where we have introduced the dimensionless phase velocity: $\underline{c} = c/U$. Here a_2 is the water wave amplitude, while a_1 is the amplitude at the wind interface.

C.2 Amplitude Ratio for Short Waves

We introduce the shortwave equation for the phase velocity from Sect. 4.3 in Chap. 3 by $c_r = (U + \varepsilon U)/2$ and $c_i = \Delta U/2$. The amplitude ratio above may now be developed as follows.

H. Moshagen, *The General Kelvin-Helmholtz Stability Model*, https://doi.org/10.1007/978-3-031-61984-7

$$\underline{c} - 1 = (1 + \varepsilon)/2 + i(1 - \varepsilon)/2 - 1 = (\varepsilon - 1)(1 - i)/2 \tag{C.4}$$

and

$$\underline{c} - \varepsilon = (1 + \varepsilon)/2 + i(1 - \varepsilon)/2 - \varepsilon = (\varepsilon - 1)(1 + i)/2 \tag{C.5}$$

The squared ratio can now be developed into

$$(\underline{c} - 1)^2/(\underline{c} - \varepsilon)^2 = (1 - i)^2/(1 + i)^2 = -1 \tag{C.6}$$

Introducing this in Eq. (C.3), we get

$$a_2/a_1 = (-1)\sinh\, kh + \cosh\, kh = \exp(-kh) \tag{C.7}$$

Thus, the relation between the amplitudes for short waves is given by the equation:

$$a_2 = \exp(-kh)a_1 \tag{C.8}$$

C.3 Amplitude Ratio for Long Waves

Equation (C.2) may be rewritten for long waves, $kh \ll 1$, as

$$a_2/a_1 = kh(\underline{c} - 1)^2/(\underline{c} - \varepsilon)^2 + 1 \tag{C.9}$$

where we have applied the definition $\underline{c} = c/U$ as well as the long wave approximations: $\sinh kh = kh$ and $\cosh kh = 1$. We introduce c from Eq. (3.23) and (3.24):

$$c_r = \varepsilon U \text{ and } c_i = U(1 - \varepsilon)(kh)^{1/2} \tag{C.10}$$

Hereby we get:

$$\underline{c} - 1 = -(1 - \varepsilon) + i(1 - \varepsilon)(kh)^{1/2} \tag{C.11}$$

and

$$\underline{c} - \varepsilon = \varepsilon - \varepsilon + i(1 - \varepsilon)(kh)^{1/2} = i(1 - \varepsilon)(kh)^{1/2} \tag{C.12}$$

The ratio between these two is now developed into:

$$(\underline{c}-1)/(\underline{c}-\varepsilon)=\left\{-1+i(kh)^{1/2}\right\}/\left(i\,(kh)^{1/2}\right) \tag{C.13}$$

By squaring this, we get the following equation:

$$(\underline{c}-1)^2/(\underline{c}-\varepsilon)^2=-\left\{1-i\,2(kh)^{1/2}-kh\right\}/kh \tag{C.14}$$

Introducing this in Eq. (C.9), we have determined the amplitude ratio as follows:

$$a_2/a_1=kh(-1)\left\{1-i2(kh)^{1/2}-kh\right\}/kh+1 \tag{C.15}$$

The last term—kh in the main parenthesis is small by the long wave assumption $kh \ll 1$ and may be disregarded as a first approximation. The final equation relating the longwave amplitudes now reads:

$$\begin{aligned} a_2 &= 2(kh)^{1/2} i\, a_1 + O(kh) \\ &= 2(kh)^{1/2}\exp(i\pi/2)\, a_1 + O(kh) \end{aligned} \tag{C.16}$$

Index

H. Moshagen, *The General Kelvin-Helmholtz Stability Model*,
https://doi.org/10.1007/978-3-031-61984-7

The manufacturer's authorised representative in the EU is Springer Nature Customer Service Centre GmbH, Europaplatz 3, 69115 Heidelberg, Germany. If you have any concerns regarding our products, please contact ProductSafety@springernature.com

Printed and bound by CPI Group (UK) Ltd, Croydon, CR0 4YY
01/10/2026
02225916-0002